Tools for High Performance Computing 2014

Christoph Niethammer • José Gracia •
Andreas Knüpfer • Michael M. Resch •
Wolfgang E. Nagel

Editors

Tools for High Performance Computing 2014

Proceedings of the 8th International
Workshop on Parallel Tools for High
Performance Computing, October 2014,
HLRS, Stuttgart, Germany

 Springer

Editors

Christoph Niethammer
Höchstleistungszentrum
Stuttgart (HLRS)
Universität Stuttgart
Stuttgart, Germany

José Gracia
Höchstleistungszentrum
Stuttgart (HLRS)
Universität Stuttgart
Stuttgart, Germany

Andreas Knüpfer
Zentrum für Informationsdienste und
Hochleistungsrechnen (ZIH)
Technische Universität Dresden
Dresden, Germany

Michael M. Resch
Höchstleistungszentrum
Stuttgart (HLRS)
Universität Stuttgart
Stuttgart, Germany

Wolfgang E. Nagel
Zentrum für Informationsdienste und
Hochleistungsrechnen (ZIH)
Technische Universität Dresden
Dresden, Germany

Cover front figure: Simulation of air flow around a cyclist. Data and Illustration by Thomas Obst. High Performance Computing Center Stuttgart (HLRS), Stuttgart, Germany

ISBN 978-3-319-16011-5   ISBN 978-3-319-16012-2   (eBook)
DOI 10.1007/978-3-319-16012-2

Library of Congress Control Number: 2015941290

Mathematics Subject Classification (2010): 68M20, 65Y20, 65Y05

Springer Cham Heidelberg New York Dordrecht London

Printed on acid-free paper

Springer International Publishing AG Switzerland is part of Springer Science+Business Media (www.springer.com)

Preface

Today advances in computer hardware are no more driven by increased operation frequencies. They are driven by parallelism and specialised accelerating features—whether by SIMD registers and compute units inside traditional CPU cores or by additional accelerator cards. In the field of high-performance computing (HPC), up to multiple thousands of these compute units can be combined to one big system via sophisticated high-speed networks nowadays to solve large computational problems within a reasonable time frame.

These advances steadily allow to solve more complex and even new problems. But it comes with the cost of complexity of the hardware, which increases the necessary programming effort to utilise the entire potential of these systems. Application developers have to write codes which can cope with parallelism keeping millions of processing elements running today. Therefore, extracting parallelism and efficient load balancing are becoming more and more an issue. At the same time the memory system holding computational data is also undergoing massive changes: Up to four levels of coherent caches can be seen in today's CPUs, while accelerators often use directly programmable caches. The arising problems are absolutely nontrivial to solve as program data structures may not fit for all CPUs and accelerators. And in a heterogeneous system, this may become even more complicated.

The European Technology Platform for High Performance Computing, an industry-led forum, emphasises in its 2013 Strategic Research Agenda the importance of the programming environment—including tools for performance analysis, debugging and automatic performance tuning—for the development of efficient, massively parallel and energy-efficient applications. This shows how vital tools in the field of HPC are not only for researchers in the academic field but also for more and more industrial users.

While traditional parallel programming models for distributed and shared memory systems as well as accelerators exist and are already standardised—just mentioning the Message Passing Interface (MPI) and Open Multiprocessing (OpenMP) API—they may not be easy to use for or even capable to exploit the potential of today's and upcoming HPC systems. So besides the traditional standards, new parallel programming models are developed—just mentioning dependency-driven task-based programming models like OmpSs or new functional programming approaches. These models may not only help programmers to ease the work of writing parallel programs but also allow the development of tools which can support them with the task of program parallelisation.

Since 2007 the International Parallel Tools Workshop provides once a year the opportunity for the leading HPC tool developers worldwide to exchange their experiences in optimisation techniques and development approaches. It covers the state-of-the-art development of parallel programming tools, ranging from debugging over performance analysis to fully and semi-automatic tuning tools as well as best practices in integrated developing environments for parallel platforms. The workshop is jointly organised by the High Performance Computing Center Stuttgart (HLRS)[1] and the Center for Information Services and High Performance Computing of the University of Dresden (ZIH-TUD).[2]

This book comprises a continuation of a successful series of publications that started with the first tools workshop in 2007. It contains contributed papers presented at the 8th International Parallel Tools Workshop,[3] held 1–2 October 2014 in Stuttgart, Germany. More than ten different tools covered different aspects of the software optimisation process and global community developments.

So, the Scalasca tool developers reported on their effort to bring their tool to the community instrumentation and measurement infrastructure Score-P—which was introduced to reduce tool development and maintenance effort for different HPC platforms itself. Allinea MAP was extended to support Hybrid MPI + OpenMP codes without adding large overheads and also added new power-usage metrics in the sense of green IT. Two tools introduced their approaches to help developers with parallelising serial programs: While DiscoPoP uses a bottom-up approach starting with compute units at the instruction level, Tareador uses an opposed approach refining tasks from code blocks until it explores enough parallelism for the target system.

Different tool developers reported on their progress in analysing large applications by aggregating and filtering important performance parameters and events. New multidimensional data aggregation methods were shown and evaluated for the trace analyser Ocelotl. A new approach using tracking analysis techniques to study performance characteristics of applications was presented, and trace and sampling data were combined to improve the detail level of performance data for large parallel programs while not increasing runtime and performance data sets too much at the same time. The effects of heterogeneous compute environments with accelerators were analysed by critical-blame analysis on the basis of the Score-P workflow, and program and runtime parameter tuning techniques were presented using the Periscope Tuning Framework. And finally a new idea of multidomain

---

[1] http://www.hlrs.de

[2] http://tu-dresden.de/die_tu_dresden/zentrale_einrichtungen/zih/

[3] http://toolsworkshop.hlrs.de/2014/

performance analysis was introduced where performance data may be mapped not only to hardware topology or features but also to actual simulation data.

We appreciate the interesting contributions and fruitful conversations during the workshop from the speakers and all participants. Special thanks go to Steffen Brinkmann and Mathias Nachtmann for their support organising the workshop.

Stuttgart, Germany   Christoph Niethammer
January 2015   José Gracia
                                                                      Andreas Knüpfer
                                                                    Michael M. Resch
                                                                    Wolfgang E. Nagel

Contents

Scalasca v2: Back to the Future

Ilya Zhukov, Christian Feld, Markus Geimer, Michael Knobloch, Bernd Mohr, and Pavel Saviankou

Abstract Scalasca is a well-established open-source toolset that supports the performance optimization of parallel programs by measuring and analyzing their runtime behavior. The analysis identifies potential performance bottlenecks – in particular those concerning communication and synchronization – and offers guidance in exploring their causes. The latest Scalasca v2 release series is based on the community instrumentation and measurement infrastructure Score-P, which is jointly developed by a consortium of partners from Germany and the US. This significantly improves interoperability with other performance analysis tool suites such as Vampir and TAU due to the usage of the two common data formats CUBE4 for profiles and the Open Trace Format 2 (OTF2) for event trace data. This paper will showcase recent as well as ongoing enhancements, such as support for additional platforms (K computer, Intel Xeon Phi) and programming models (POSIX threads, MPI-3, OpenMP4), and features like the critical-path analysis. It also summarizes the steps necessary for users to migrate from Scalasca v1 to Scalasca v2.

1 Motivation and Introduction

The Scalasca toolset [10, 23] is a portable, well-established software package which supports the performance optimization of parallel applications by measuring and analyzing their dynamic runtime behavior. It has been specifically designed for use on large-scale systems including IBM Blue Gene and Cray X series, but is also well-suited for small- and medium-scale HPC platforms. A distinctive feature of Scalasca is its scalable automatic trace analysis, which identifies potential performance bottlenecks – in particular those concerning communication and synchronization – and offers guidance in exploring their causes. Scalasca is available under a 3-clause BSD open-source license.

I. Zhukov • C. Feld • M. Geimer • M. Knobloch • B. Mohr (✉) • P. Saviankou
Forschungszentrum Jülich GmbH, Jülich Supercomputing Centre
52425 Jülich, Germany
e-mail: i.zhukov@fz-juelich.de; c.feld@fz-juelich.de; m.geimer@fz-juelich.de;
m.knobloch@fz-juelich.de; b.mohr@fz-juelich.de; p.saviankou@fz-juelich.de

© Springer International Publishing Switzerland 2015   1
C. Niethammer et al. (eds.), *Tools for High Performance Computing 2014*,
DOI 10.1007/978-3-319-16012-2_1

Users of the Scalasca toolset can choose between two different analysis modes: (i) performance overview on the call-path level via profiling and (ii) the analysis of wait-state formation via event tracing. Wait states often occur in the wake of load imbalance and are serious obstacles to achieving satisfactory performance and scalability. The latest release at the time of writing (v2.1) also includes a scalable critical-path analysis [3]. Analysis results are presented to the user in an interactive explorer called Cube that allows the investigation of the performance behavior on different levels of granularity along the dimensions metric, call path, and process/thread. Selecting a metric displays the distribution of the corresponding value across the call tree. Selecting a call path again shows the distribution of the associated value over the machine or application topology. Expanding and collapsing metric or call tree nodes allows to investigate the performance at varying levels of detail.

Scalasca and Cube are both under active development. New features – in many cases triggered by user requests – are developed, tested, and integrated into the main development branch. In addition, bugs that emerge are fixed and performance as well as scalability are constantly evaluated and improved. The latest Scalasca v2 release series is based on the community-driven instrumentation and measurement infrastructure Score-P [15], which is jointly developed by a consortium of partners from Germany and the US. This significantly improves interoperability with other performance analysis tool suites such as Vampir [14] and TAU [25], enabled by the usage of two new common data formats: CUBE4 for profiles and the Open Trace Format 2 (OTF2) for event trace data.

Re-architecting Scalasca on top of Score-P marked an important milestone for the Scalasca project. While the previous Scalasca v1 release series provided a rather monolithic, mostly self-contained toolset which included all components required to perform measurements and performance analyses of parallel applications in a single package, the Score-P project followed a component-oriented philosophy from the very beginning – with the intention to allow a more decoupled development of the individual components and to share them among several tools. Therefore, the changeover to Score-P naturally lend itself to also "componentize" the Scalasca toolset.

From a user's perspective, the most obvious difference between the Scalasca v1 series and Scalasca v2 is that one now has to install several packages instead of just one. In terms of actual usage, however, we tried to keep the convenience commands and their options as stable as possible, although a few changes were necessary to adapt to new features and to provide new functionality. Nevertheless, we believe that the benefits of the long-term commitment of the Score-P partners and the increased tool interoperability overall far outweigh the additional (one-time) learning effort.

The remainder of this article is structured as follows: In Sect. 2, we briefly introduce the community instrumentation and measurement system Score-P. Section 3 then describes the performance analysis toolset Scalasca, and outlines changes and differences between the old (v1) and new (v2) generation of Scalasca in more detail. Finally, Sect. 4 gives an overview of new features introduced in the latest versions of Score-P, Scalasca, and Cube, as well as future plans for enhancements.

2   Scalable Performance Measurement Infrastructure for Parallel Codes (Score-P)

The community performance measurement and runtime infrastructure *Score-P* [15] – Scalable Performance Measurement Infrastructure for Parallel Codes – is a highly scalable and easy-to-use tool suite for profiling, event tracing, and online analysis of HPC applications.

Until 2009, the performance tools landscape was fragmented and revealed many redundancies – not so much in the analysis parts, but in program instrumentation, measurement data acquisition, and data formats. The analysis tools Scalasca [10], Vampir [14, 27], TAU [25], and Periscope [1] all provided their native instrumentation and measurement systems, data formats, and configuration options. Although very similar in functionality, the effort to maintain a portable and scalable measurement tool with associated data formats was redundantly expended by every tool developer group. Moreover, on the user's side, dealing with different measurement tools that expose incompatible configurations, diverse feature sets, and different options for equivalent features, created discomfort and made interoperability between analysis tools difficult.

Therefore, a consortium of tool developer groups from Germany and the US as well as other stakeholders in 2009 started the Score-P project to remedy this situation. Score-P aims to provide a single, highly scalable, and easy-to-use tool suite for profiling, event tracing, and online analysis of HPC applications written in C/C++ and Fortran. Score-P comes along with scalable, space-efficient, and well-defined data formats: CUBE4 for profiling and OTF2 [7] for tracing data. Based on the experience with the existing tools, the interaction with their users, and the programming paradigms most prevalent in HPC, the Score-P architecture as shown in Fig. 1 was derived.

The workflow to obtain measurement data usually consists of instrumentation, measurement configuration, and running the instrumented application. At instrumentation time, this typically means to recompile/relink the application using the `scorep` compiler wrapper as a prefix to the actual compile/link commands. This step inserts measurement probes into the application code that collect performance-related data when triggered during measurement runs. Currently, the following means of instrumentation are supported:

- Compiler-based instrumentation (entry/exit of functions)
- MPI/SHMEM library interposition
- OpenMP source-code instrumentation using OPARI2
- POSIX thread instrumentation using library wrapping
- CUDA event interception using the CUPTI library
- Source-code instrumentation via the TAU instrumentor [9]
- User instrumentation using convenient macros

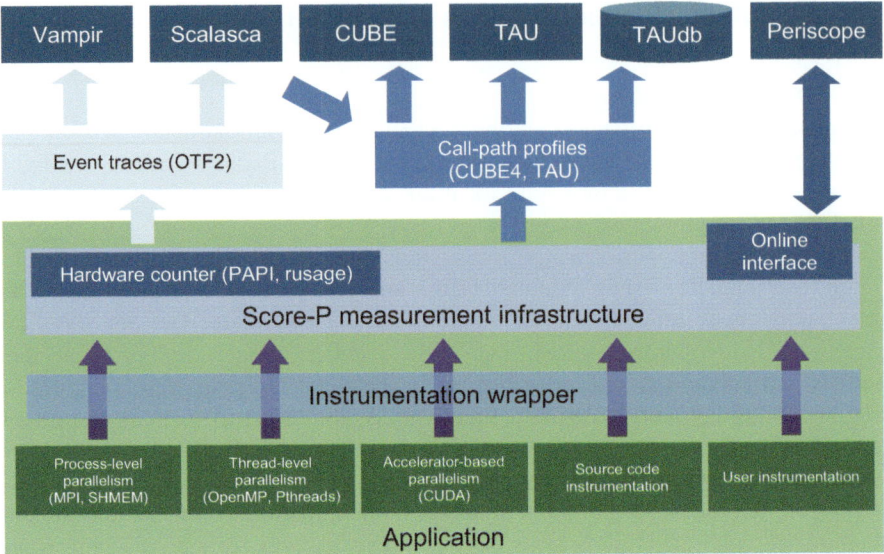

Fig. 1 Score-P architecture overview. Instrumentation wrappers capture important events during application execution and pass them to the Score-P measurement core. Here, this data is optionally combined with hardware counter information, and detailed event traces or call-path profiles are generated, which can be handled by various analysis tools. In addition, an online interface allows tools to access profiling data already during measurement

The instrumentation probes are placed at performance-relevant regions in the code, e.g., function entry and exit, thread-management functions, and communication routines. Once triggered, the probes obtain relevant performance data and pass them to the Score-P measurement core where they are augmented with timestamps and, optionally, additional metrics from hardware and OS counter sources. This enhanced event data is then processed by so-called substrates, currently either profiling and/or tracing. Substrates collect and process the measurement events thread-locally and write the results efficiently to file. The profiling substrate writes either CUBE4 or TAU profiles, while the tracing substrate outputs its data in the space-efficient OTF2 format.

Measurements can be conveniently configured using environment variables. That is, once instrumented, the application can be run in different measurement configurations without recompilation. Depending on the settings of the configuration variables, users can

- Switch between profiling and tracing,
- Apply event filters to control overhead,
- Collect additional metrics (e.g., hardware counters),
- Adapt measurement internals (e.g., size of memory buffers),

- Specify output parameters,
- And much more.[1]

After its first public release end of 2011, the Score-P infrastructure has become the default instrumentation and measurement system for Scalasca v2, Vampir, and Periscope, and is also supported by the TAU performance system as a configuration option. It is actively maintained and constantly enhanced with new features by the contributing partners, managed by a meritocratic governance model.

3   Scalasca v2

As already mentioned in the introduction, the changeover to the Score-P instrumentation and measurement system as the new foundation of the Scalasca toolset marked an important milestone for the Scalasca project. While the previous Scalasca v1 release series provided all components required to perform measurements and performance analyses of parallel applications in a single distribution package, the Score-P project follows a rather different philosophy. Here, smaller components are developed and released in a more decoupled fashion – with the intention to share them among several tools. Therefore, switching to Score-P naturally lend itself to also "componentize" the Scalasca toolset, eliminate some legacy components, and substitute a number of previously internal components by now external components from the Score-P universe:

- The custom instrumentation and measurement system of Scalasca v1 (EPIK) was replaced by Score-P.
- The libraries to read/write Scalasca's original custom trace file format (EPILOG) were substituted by OTF2.[2]
- The graphical report explorer Cube v3 as well as associated read/write libraries and command-line tools were split off as a separate component, significantly improved, and subsequently released in the Cube v4 package.
- The legacy serial trace analyzer from Scalasca's predecessor project KOJAK as well as the non-scalable trace converters from the EPILOG to the Jumpshot and Paraver trace formats were removed.

These substantial changes, however, posed a number of challenges for both the Scalasca developers and the Scalasca user community.

---

[1]The command `scorep-info config-vars` provides an exhaustive list of options available for a particular installation.

[2]A stripped-down EPILOG reader library is still included in Scalasca v2 to provide backwards compatibility support for existing trace files, except for traces stored in SIONlib [8] containers which are not supported.

From a developer's point of view, adjusting the code base of the remaining core components of Scalasca (i.e., its scalable trace-based analysis tools) required a significant development effort. On the one hand, code sections dealing with file I/O had to be reworked, since the APIs of the Cube v4 and OTF2 libraries differ significantly from their previous counterparts. Moreover, both Cube and OTF2 are under continuous development and now provided as external components, thus creating a need to identify – and potentially deal with – different versions of these libraries. On the other hand, OTF2 also introduced many small differences in the trace definition and event data compared to the previous EPILOG format, which required a careful inspection and adaption of Scalasca's trace analysis algorithms.

From a user's perspective, the most obvious difference between the Scalasca v1 series and Scalasca v2 is that now several packages have to be installed instead of just one. In terms of actual usage, however, we tried to keep the convenience commands (scalasca and its shortcuts skin, scan, and square) and their command-line options as stable as possible. Nevertheless, some changes were necessary, which will be described in more detail below. All in all, however, we believe that the benefits of the improved functionality, the long-term commitment of the Score-P partners, and the increased tool interoperability far outweigh the additional installation and (one-time) learning effort.

3.1   General Comparison of Scalasca v1 and Scalasca v2

As mentioned above, basically all components of Scalasca v2 had to be adjusted and/or refactored to integrate with the Score-P instrumentation and measurement system. In the following, however, we focus on the user-visible differences. That is, we compare the usage of the two generations of the Scalasca toolset to help existing users with transitioning from the old to the new version more easily. An initial comparison of their performance and scalability can be found in [28].

Table 1 summarizes the differences between Scalasca v1 and Scalasca v2 from a high-level perspective. While at first glance the table suggests that there are – except for the license – no commonalities between the two generations of Scalasca, the detailed comparisons in the following subsections will show that the differences are often, especially with respect to the Scalasca convenience commands, only marginal and in most cases motivated by improved functionality provided by Score-P.

Obviously, some usability changes can barely be avoided when replacing a core component such as the instrumentation and measurement infrastructure by a unified solution that emerged from existing measurement systems of multiple tool suites. For example, Score-P's instrumenter command uses different command-line options than the instrumenter provided by Scalasca's former measurement system EPIK. Moreover, their manual instrumentation APIs are incompatible, though the API supported by Score-P is much more feature rich. In addition, the names (and often also the valid values) of the environment variables used to configure measurements

Table 1 High-level comparison of Scalasca v1 and Scalasca v2

	Scalasca v1	Scalasca v2
Instrumentation system	EPIK	Score-P
Command line switches	Different	
Manual instrumentation API	Different	
Environment variables	Different	
Filter file syntax	Different	
Memory buffers	separate for each thread	memory pool on each process
Experiment directory name	`epik_###`	`scorep_###`
Profile format	CUBE3	CUBE4
Trace format	EPILOG	OTF2
Scalable trace I/O	Supports SIONlib	Partially supports SIONlib
License	3-clause BSD	

have changed. Therefore, it is inevitable that existing users of Scalasca transitioning to the new version have to familiarize themselves with these modifications.

Another notable change concerns the syntax of filter files, which are commonly used to control the measurement overhead by excluding source-code regions from being measured. While Scalasca v1 only supported a simple plain text file listing function names (optionally using wildcards) that should be ignored during measurement, Score-P provides the ability to filter based on function names as well as file names, also supporting wildcards. In addition, both blacklisting and whitelisting of functions/files are now supported. This, however, necessitated the introduction of an incompatible, more powerful syntax for filter files.

Finally, handling of memory buffers internal to the measurement system has changed. Scalasca v1 provided two environment variables to control its memory usage: one to define the size of a per-process buffer for definition data and another to set the size of a per-thread trace data buffer, the latter only if tracing mode is configured. By contrast, Score-P uses a memory pool on each process, whose total size can be controlled using a single variable. This approach is typically more memory efficient and in general easier to use.

Other differences such as the experiment directory prefix and the different profile and trace file formats used by the two generations of Scalasca, although apparent, usually do not affect the user directly. One notable exception are the different levels of support for writing trace data in SIONlib [8] containers. At the time of writing, the current Score-P release (v1.3) only supports SIONlib containers for OTF2 traces from pure MPI applications, while the former EPIK measurement system also included support for hybrid MPI+OpenMP codes. This currently limits the scalability of tracing in Score-P for such applications. However, extending OTF2 and Score-P to support SIONlib for all supported programming models (and their combination) is already work in progress.

3.2   Changes in the Scalasca Convenience Commands

Although there are many differences between Scalasca v1 and Scalasca v2, we
tried to preserve the overall usage workflow and the set of associated convenience
commands. Therefore, users already familiar with Scalasca v1 should be able to
continue using these commands to accomplish the most common tasks.

Most of Scalasca's functionality can be easily accessed through the `scalasca`
command, which provides action options that in turn invoke the corresponding
underlying commands: `scorep` for instrumentation, `scan` for measurement execution control, and `square` for analysis report postprocessing and examination.
These actions are:

- `scorep`
  is used to instrument the application during compilation/linking. The former
  `scalasca -instrument` (or short `skin`) command is now deprecated and
  only provided to assist in converting existing measurement configurations to
  Score-P. Hence, it tries to map the command-line options of the Scalasca v1
  instrumenter onto corresponding options of the `scorep` command as far as this
  is possible, and prints the new command to standard output. However, to take
  full advantage of Score-P's functionality, it is recommended to use the `scorep`
  instrumenter command directly.
- `scalasca -analyze` (or short `scan`)
  is used to control the Score-P measurement environment during the execution
  of the target application, and to automatically initiate Scalasca's trace analysis
  after measurement completion in case tracing was requested.
- `scalasca -examine` (or short `square`)
  is used to postprocess the analysis report generated by a Score-P profiling
  measurement and/or Scalasca's automatic post-mortem trace analysis, and to
  start the analysis report examination browser Cube.

Each of these commands is discussed in further detail below.

3.2.1   Scalasca Application Instrumenter

In order to perform measurements with Scalasca, the application's source code
has to be prepared, a process called instrumentation. That is, special calls have
to be inserted into the code which trigger measurements at performance-relevant
points during application runtime. The Score-P measurement system can do this
instrumentation in various ways; the most important ones are listed below.

Compiler instrumentation inserts specific instrumentation probes at the enter
and exit of functions. This method is supported by most modern compilers and
can usually be enabled by a particular compiler flag. It is the most convenient
way to instrument an application but may result in high overhead and/or disruptive

measurements. Both of these issues can be addressed by manual instrumentation and/or measurement filtering (see Sect. 3.2.2).[3]

If compiler instrumentation is not possible or it incurs too much overhead, one can use *manual instrumentation*. Here, the user manually inserts function calls into the application's source code, using the provided manual instrumentation API. Note that the Score-P API differs from the previous EPIK API. Besides changing the common prefix of the API calls from EPIK to SCOREP, the calls themselves have been renamed to be more expressive and now also take different parameter lists. This, however, allowed to make the API more flexible and powerful, for example, to support phase and dynamic region profiling. For more detailed information on the new possibilities, please consult the Score-P user manual [24].

Compiler-based and manual instrumentation can also be used in combination, thereby allowing to mark specific regions other than function entry and exit and thus augmenting the measurement output with additional information. While manual instrumentation is very flexible, it is more time-consuming and requires extra implementation efforts.

If Score-P has been configured with support for the Program Database Toolkit (PDToolkit) [9], *automatic source-code instrumentation* can be used as an alternative to compiler instrumentation. In this case, the source code of the target application is pre-processed before compilation and appropriate calls to the manual instrumentation API are inserted automatically. PDT instrumentation allows for instrumentation-time filtering, thus potentially reducing overhead.

In the rare case that neither automatic compiler instrumentation nor automatic source-code instrumentation via PDToolkit are supported, the user can apply *semi-automatic instrumentation*. Here, instrumentation can be accomplished by means of the Open Pragma And Region Instrumenter (OPARI2) [20], a source-to-source instrumenter that was originally developed to detect and instrument OpenMP [21] directives. OPARI2 scans the source code for OpenMP directives, inserting specific instrumentation probes according to the POMP2 interface [18]. The same procedure can be used to instrument user regions defined via pragmas or to instrument directive-based non-OpenMP paradigms [13]. The advantage of this method is that the pragma-based instrumentation directives will be ignored by the compiler during "normal" compilation.

Table 2 shows a comparison between the most relevant options that can be passed to the old and the new instrumentation commands. As can be seen, the command-line options supported by the scorep command are very similar to those used by scalasca -instrument or skin, respectively. However, the Score-P instrumenter command provides many additional options not listed in the table, making it more flexible and feature rich than its predecessor.[4]

---

[3]Some compilers, e.g., Intel, provide command-line options to control instrumentation-time filtering, however the functionality provided is typically limited and very compiler (and compiler version) specific.

[4]The command scorep --help provides an exhaustive list of instrumentation options.

Table 2 Comparison between the `skin` (`scalasca -instrument`) and `scorep` commands

Scalasca v1		Scalasca v2		
Command	Option	Command	Option	Description
skin	-v	scorep	-v	Enable verbose mode
			--verbose=<value>	
	-comp=all		--compiler	Turn on compiler instrumentation (default)
	-comp=none		--nocompiler	Turn off compiler instrumentation
	-user		--user	Turn on user instrumentation
	-pdt		--pdt	Process source files with PDT instrumenter
	-pomp		--pomp	Process source files for POMP directives
			--dry-run	Display executed commands. No execution
			--keep-files	Do not delete temporarily created files

3.2.2 Scalasca Measurement Collection and Analysis Nexus

The Scalasca measurement and analysis nexus (`scalasca -analyze` or short `scan`) configures and manages the collection of performance experiments of an application instrumented by Score-P. By setting different measurement and analysis configurations, users can perform various experiments using a single instrumented executable without re-instrumentation. Typically, a single performance experiment is collected in an unique archive directory that contains not only measurement and analysis reports but also a measurement log file and configuration data. Score-P uses a directory name prefix of `scorep_` for this directory, while Scalasca's EPIK measurement system used `epik_`.

Users can choose between generating a summary analysis report (profile) with aggregate performance metrics for each function call path and/or generating per-thread event traces recording time-ordered runtime events which are automatically fed to Scalasca's trace analyzer to identify potential communication and performance bottlenecks. Alternatively, traces can be used for timeline visualization using, for example, the Vampir trace browser.

Summarization is particularly useful to get a first insight into the performance behavior of the inspected application as well as to optimize the measurement configuration for subsequent trace generation. If hardware counter metrics were requested, these are also included in the summary report. By contrast, tracing provides detailed insights into the dynamic application behavior.

Measurement overhead can be prohibitive for small, frequently-called routines. Therefore both Scalasca v1 and Score-P provide measurement filtering capabilities.

To enable filtering, the user has to prepare a filter file. Scalasca v1 supports a plain text file containing the names of functions to be excluded from measurement, whereas Score-P can filter based on region name *and* source file names. While the use of wildcards is allowed by both file formats, only Score-P supports white- and blacklisting – also in combination. The enhanced filtering mechanism in Score-P requires a more powerful filtering syntax which will be explained in the following paragraph.

Score-P's filter file can contain the following two sections:

- A section with rules for *region names*. This section must be enclosed by the keywords SCOREP_REGION_NAMES_BEGIN and SCOREP_REGION_NAMES_END.
- A section with rules for filtering all regions defined in specific *source files*. This section must be enclosed by the keywords SCOREP_FILE_NAMES_BEGIN and SCOREP_FILE_NAMES_END.

Within these sections, the user can specify an arbitrary number of include and exclude rules which are processed in sequential order. An exclude rule starts with the keyword EXCLUDE whereas include rules start with INCLUDE, both followed by one or several white-space separated file or region names. Bash-like wildcards can be used in file or/and region names.

By default, all files and regions are included. After all rules of the filter file have been evaluated, files and regions that were marked as excluded are filtered, i.e., excluded from measurement.

Besides the two filter sections, users may also use comments in the filter file. Comments start with the '#' character and are effective throughout the entire rest of the line. If a region name or source file name contains the comment character '#', the user must escape it with a backslash.

Let us consider the following filter file:

```
#Exclude all regions except bar and foo
SCOREP_REGION_NAMES_BEGIN
     EXCLUDE *
     INCLUDE foo bar
SCOREP_REGION_NAMES_END
#Exclude all files except *.c
SCOREP_FILE_NAMES_BEGIN
     EXCLUDE *
     INCLUDE *.c
SCOREP_FILE_NAMES_END
```

According to this filter file only foo and bar region names in *.c source files will be recorded by the measurement system.

To collect measurements for profiling or/and tracing, additional memory for internal data structures as well as for the trace data is required. Even more memory is needed if hardware counter measurements are requested. This additional memory in Scalasca v1 and Score-P constitutes the memory buffer. Although memory buffers are used in both measurement systems, their handling changed notably. Scalasca v1 uses separate pre-allocated memory buffers for each thread whereas Score-P uses a pre-allocated memory pool consisting of a set of chunks on each process. Score-P's memory pool dynamically distributes memory chunks to threads that demand memory to store event data.[5] This approach manages memory in a more flexible but still efficient way and thus minimizes overhead due to run-time memory allocations.

If tracing is enabled, each thread will generate a trace file containing chronologically ordered local events. However, for applications running at scale the *one file per thread* approach might impose a serious I/O bottleneck. To address this issue Score-P can leverage the SIONlib I/O library. SIONlib improves file handling by transparently mapping logical thread-local files into a smaller number of physical files.

Once a measurement run terminates, the Scalasca trace analysis is automatically initiated to quantify wait states that cannot be determined with runtime summarization. The report generated by the trace analysis is similar to the summary report but includes additional communication and synchronization inefficiency metrics.

Table 3 shows a comparison of the most relevant options and environment variables[6] that can be passed to the two different versions of the scalasca -analyze or scan command that controls measurement collection and analysis. As can be seen, the command-line options remained unchanged. By contrast, the names of all environment variables have been changed in Score-P. Score-P provides more self-explanatory variable names prefixed with SCOREP_ rather than with EPIK_. We believe that these names can be more easily memorized by users than their previous counterparts.

3.2.3 Scalasca Analysis Report Explorer

Once measurement and analysis are completed, analysis reports are produced in either CUBE3 (Scalasca v1) or CUBE4 (Scalasca v2) format, which can be interactively explored with the Cube GUI and processed by various Cube command-line utilities. Although the CUBE format has been changed considerably with the new generation of Scalasca, the Cube GUI and utilities provide the same functionality as in the previous version.

In the usual workflow, the scalasca -examine (or short square) convenience command will post-process intermediate analysis reports produced by the measurement and analysis to further derive additional metrics and construct a hierarchy of measured and derived metrics. The Cube browser will then present the final

---

[5]The size of the chunk, a so-called page, is configurable using the SCOREP_PAGE_SIZE environment variable. The other memory-related configuration variable is SCOREP_TOTAL_MEMORY, denoting the total amount of memory reserved by each process.

[6]For Score-P, the command scorep-info config-vars provides the list of all environment variables (and their default values) that are available for the current installation.

Table 3 Comparison of the `scalasca -analyze` (`scan`) command-line options and the most important, associated environment variables

Command switch	Environment variable [default value]		Description
	Scalasca v1	Scalasca v2	
-s	EPK_SUMMARY [1]	SCOREP_ENABLE_ PROFILING [1]	Enable runtime summarization
-t	EPK_TRACING [0]	SCOREP_ENABLE_ TRACING [0]	Enable trace collection and analysis
-e <dir_name>	EPK_TITLE	SCOREP_EXPERIMENT_ DIRECTORY	Experiment archive to generate and/or analyze
-f <filter_file>	EPK_FILTER	SCOREP_FILTERING_ FILE	File specifying measurement filter
	ESD_BUFFER_SIZE [100000], ELG_BUFFER_SIZE [10000000]	SCOREP_TOTAL_ MEMORY [16384000]	Total memory in bytes for the measurement system
	EPK_METRICS[a]	SCOREP_METRIC_ PAPI[b]	List of counter metrics
	ELG_SION_FILES [0]	SCOREP_TRACING_ MAX_PROCS_PER_ SION_FILE [1024]	Maximum number of processes that share one sion file

[a] metrics provided as a colon-separated list
[b] by default metrics provided as a comma-separated list (separator can be configured with *SCOREP_METRIC_PAPI_SEP* environment variable)

report. The command-line options supported by `scalasca -examine/square` did not change during the transition from Scalasca v1 to Scalasca v2. Therefore users can continue to work with already familiar commands and command-line options during this stage of the workflow.

4 New Features and Future Plans

Score-P, Scalasca and Cube are under active development. New features, also on user request, are developed, tested, and integrated into the main development branch. Bugs that emerge are fixed and performance is constantly evaluated and improved. In this section, we highlight recently added features and improvements, and outline our plans for future work.

4.1 Score-P

In collaboration with our Score-P development partners

- Technische Universität Dresden,
- Technische Universität München,

- RWTH Aachen University,
- German Research School for Simulation Sciences (Aachen), and
- University of Oregon

we provide a small number of feature releases each year. Each feature release is usually followed by one or more bugfix releases. We do not follow a fixed release schedule but ship a new version if one or more user-relevant features become available.

4.1.1 Score-P v1.3

The last feature release was Score-P v1.3, released August 2014. Significant new features in Score-P v1.3 include:

- Basic support for the K Computer and Fujitsu FX10 systems. Fujitsu systems are cross-compile systems using rarely encountered Fujitsu compilers. The porting effort was significant, but as Score-P is a joint infrastructure, it had to be done just once.[7]
- Support for instrumenting programs which use Symmetric Hierarchical Memory access (SHMEM) library calls for one-sided communication. SHMEM implementations from Cray, Open MPI, OpenSHMEM, and SGI are supported.
- Basic support for POSIX threads instrumentation. At the moment Score-P is capable of tracking creation and joining of POSIX threads as well as important synchronization routines.[8]
- Support for CUDA [19] versions 5.5 and 6.0.
- Improved event size estimation in `scorep-score` to adapt memory requirements for subsequent trace experiments.

---

[7]The Tofu network topology will be supported in a subsequent release.

[8]Currently supported POSIX threads routines are `pthread_create`, `pthread_join`, `pthread_mutex_init`, `pthread_mutex_destroy`, `pthread_mutex_lock`, `pthread_mutex_trylock`, `pthread_mutex_unlock`, `pthread_cond_init`, `pthread_cond_destroy`, `pthread_cond_signal`, `pthread_cond_broadcast`, `pthread_cond_wait`, and `pthread_cond_timedwait`. The following thread management functions are currently not supported and will abort the program: `pthread_exit` and `pthread_cancel`. The usage of `pthread_detach` will cause the program to fail if the detached thread is still running after the end of the main routine. These limitations will be addressed in an upcoming version of Score-P. Note that currently every thread creation needs to be instrumented.

4.1.2 Upcoming Score-P v1.4

Score-P v1.4, likely to be released by the end of 2014, might include some of the features we and our project partners are currently working on. The prominent user-visible ones are:

- Basic OpenCL [26] support by intercepting OpenCL library routines using library wrapping.
- Intel Xeon Phi support for native and symmetric mode.
- GCC plugin as an alternative to the default compiler instrumentation, allowing for instrumentation-time filtering.

The next release will also include improved POSIX threads support, internal refactorings, and bugfixes.

4.1.3 Future Plans

There are a number of features we are working on that eventually will appear in a future release. Some of them are listed below:

- Experimental support for sampling as an alternative to instrumentation-based measurement, potentially combined with library wrapping and interposition, will give opportunity to collect measurements with lower overhead than with full instrumentation [12].
- Basic OpenACC [5] instrumentation using the OpenACC tools interface.
- A refactored OPARI2 allows to easily add source-to-source instrumentation directives, i.e., C/C++ pragmas and Fortran special comments, beyond the already supported OpenMP ones.
- Hardware and application-specific Cartesian topologies that were already supported by Scalasca v1 will be ported to Score-P and Scalasca v2.
- MPI-3 [17] support. MPI-3.0 is a major update to MPI adding over 100 new functions and extensions to collective and one-sided communication, which we expect to be used by scientific applications in the near future. We plan to add support for MPI-3 features in multiple steps.

  In the current release, MPI-3 functions are silently ignored by the measurement system, so first we will provide basic wrappers for all new MPI-3 functions. The next step is adding support for the newly introduced non-blocking collectives and neighborhood collectives. Non-blocking collectives are similar to non-blocking point-to-point routines, i.e., they have the same semantics as their blocking counterparts but can be used to overlap computation and communication. Neighborhood collectives on the other hand introduce a new semantic to collective operations, as only neighbors in a defined (Cartesian or distributed graph) topology communicate with each other. We will initially concentrate on Cartesian topologies and tackle graph topologies later. Concurrently, we are working on support for the new RMA features to analyze applications using one-sided communication. In all cases profiling support will be implemented first. Tracing support for these new features may require new OTF2 event records and adaptations in Scalasca (for analysis) and Vampir (for visualization).

MPI-3 also offers new Fortran 2008 bindings, which will probably be used once Fortran 2008 features are exploited by more scientific applications. Currently, all MPI wrappers are implemented in C with Fortran-to-C conversion to support the Fortran bindings in MPI-2. We will need to implement new native Fortran wrappers to support the Fortran 2008 bindings in MPI-3.

Finally, we plan to support the MPI Tools interface (MPI_T). MPI_T exposes performance (and control) variables internal to the MPI library to external tools to help the advanced user in fine-tuning the MPI usage in their application (or control the MPI library accordingly). Score-P will primarily focus on performance variables, though also control variables might be exploited by tools like Periscope.

- New programming models or extensions of existing ones:

  - Support of relevant OpenMP 4.0 [21] features, e.g., target and cancellation directives, as well as support for the OpenMP Tools Application Programming Interface for Performance Analysis (OMPT) [6].
  - Support for OmpSs [4], an extension to OpenMP to support asynchronous parallelism with data dependencies.
  - Support for Qt, ACE, and TBB threading models using Intel's dynamic binary instrumentation tool Pin [16].
  - Enhanced tasking support, in particular for MTAPI [11], using Pin as well as library wrapping.

- Basic support for the Windows platform, however, not targeting distributed-memory applications. The preferred instrumentation method will be Pin.
- New substrates – substrates are the event consumers – besides profiling and tracing. We work on a generic substrate layer that allows arbitrary processing of event data. One could think of online tuning of runtime parameters or streaming event data to reduce memory demands.
- Improved support for threading. As of Score-P v1.3, the user can either analyze programs using OpenMP *or* POSIX threads. Additionally all thread creations need to be instrumented. This is particularly challenging if the thread creation takes place in a shared library. A future version of Score-P will allow the combination of OpenMP and POSIX threading. Additionally, events from uninstrumented threads will be handled gracefully.

4.2 Scalasca v2

The first version of the Scalasca v2 series based on Score-P v1.2, OTF2 v1.2, and CUBE v4.2 was released in August 2013. Besides a new build system based on

GNU autotools, a significant amount of code refactoring, and support for the new data formats, this release included only minor functional changes and fixes for a number of bugs uncovered during the refactoring process. This initial release then served as a basis for all subsequent developments.

4.2.1 Scalasca v2.1

Scalasca v2.1, the first feature release based on the new infrastructure, was released in August 2014 and included the following new features:

- Support for the K Computer and Fujitsu FX10 systems, to bring Scalasca's platform support in line with Score-P v1.3.
- Improved detection of Late Receiver pattern instances. Previously, Late Receiver instances in non-blocking point-to-point MPI communication could only be detected in operations finalizing a single request (e.g., MPI_Wait). The enhanced detection algorithm now also supports multi-request finalization calls, such as MPI_Waitall.
- Critical-path analysis [3] for MPI- and/or OpenMP-based applications. The critical-path analysis determines the call-path profile of the application's critical path, highlighting the call paths for which optimization will prove worthwhile and thus guiding optimization efforts more precisely.

4.2.2 Scalasca v2.2

The upcoming Scalasca v2.2 release, planned for early 2015, is likely to include a number of additional features which will be briefly described below.

- Intel Xeon Phi support for native and symmetric mode.
- Basic OpenMP task support. So far, Scalasca's automatic trace analyzer has been unable to handle traces including events related to OpenMP tasking, thus not even allowing an MPI-only analysis for applications using OpenMP tasks in their computational phases. We therefore extend the analysis infrastructure to support tasks, handle (potentially) multiple call stacks, and track the asynchroneous task behavior. However, this initial tasking support will not include any task-specific analyses.
- Basic support for create/wait threading models (e.g., POSIX threads). We intend to implement some basic analyses that should work in many cases, however, the current OpenMP-based replay analysis approach used by Scalasca's trace analyzer may impose some limitations.
- In multi-threaded programs, lock contention can be a serious performance bottleneck. We are therefore investigating a more advanced analysis to identify waiting time due to lock contention, supporting both fork/join threading models (e.g., OpenMP) as well as create/wait threading models (e.g., POSIX threads).

- Finally, we plan to integrate an updated, production-ready version of the delay analysis research prototype developed for Scalasca v1 [2]. The delay analysis is capable to identify the root causes of wait states that materialize at communication or synchronization operations (i.e., the delays/imbalances causing them), and to quantify their overall costs.

4.2.3 Future Plans

In future releases, we plan to further improve the analysis capabilities of Scalasca's automatic trace analyzer in various ways:

- MPI-3 support. Currently, Scalasca's trace analyzer focuses exclusively on communication and synchronizition operations as defined by MPI-2.2. However, once MPI-3 support has been added to the Score-P measurement system, the trace analysis will be enhanced to also support MPI-3 features. In this context, non-blocking and neighborhood collective operations are of particular interest. While extending the existing trace analysis algorithms to handle non-blocking collectives should be fairly straightforward, analyzing neighborhood collectives requires enhanced wait-state detection algorithms which take the neighborhood information into account. Another area of interest is the extension of the existing Scalasca RMA analysis to support the new RMA functionality of MPI-3.
- Improved threading support. The support for create/wait thread models planned for Scalasca v2.2 is very limited and needs to be improved in various ways. For example, the timestamp correction algorithm implemented in the Scalasca trace analyzer currently only supports fork/join threading models. Thus, correcting the timestamps of an application trace using, for example, MPI and POSIX threads in combination may lead to unexpected clock condition violations due to corrections applied on the thread executing the MPI calls but not on the other threads. In addition, the critical-path and delay analyses have to be extended accordingly to support all threading models handled by the wait-state search carried out by the automatic trace analyzer.
- Another topic of active research are analyses related to task-based programming models. However, this is still preliminary work that needs thorough investigation.

4.3 Cube v4

Cube v4 and its predecessor Cube v3 both provide a graphical user interface to visually examine profile and trace analysis reports, as well as a set of libraries and tools to read, write, and manipulate such reports. Although, CUBE4 and CUBE3 reports have a very similar structure in the metric tree, call tree, and system tree, a completely new format for storing the measured data was introduced. While Cube v3 used a pure XML file format, Cube v4 uses a binary format to store the experiment data, whereas XML is only used to store metadata information. The new format provides both greater scalability and flexibility. In addition, Cube v4 introduced the following new extensions:

- A Java reader library allows TAU to import data from CUBE4 files into PerfExplorer [22] and opens the opportunity to use CUBE4 files on Android in the future.
- A flexible remapping based on specification files enables the fine-grained control of metrics and metric hierarchies that are created during remapping (e.g., for POSIX threads, CUDA, etc.).
- Support for new platforms, e.g., Windows, K Computer, and Fujitsu FX10.

In addition, we want to highlight two features we believe will benefit many users of Cube v4: derived metrics and the visualization of high-dimensional Cartesian topologies.

4.3.1 Derived Metrics in Cube

Derived metrics are a very powerful and flexible tool allowing users to define and calculate new metrics directly within Cube. While Cube's predefined metrics are stored in the analysis report, the values of derived metrics are calculated on-the-fly when necessary according to user-defined arithmetic expressions formulated using a domain-specific language called *CubePL*.[9]

A first version of *CubePL* was introduced with Cube v4.1, which had very limited functionality and allowed only simple expressions for derived metrics. The latest implementation of *CubePL* available in Cube v4.2.3 supports different kinds of derived metrics which will be explained in the following example. This version features an improved memory layout and added an extended set of predefined variables. Finally, it also provides better support for string and numeric variables, thereby allowing the user to define very sophisticated metrics.

The following simple example introduces the basic concept behind derived metrics and explains why different kinds of derived metrics are needed. Let us assume that an application with the following call tree was executed:

```
main
  - foo
  - bar
```

where *main*, *foo* and *bar* are function calls, with *foo* and *bar* being called from *main*. Further, assume that the number of floating-point operations (FLOP) has been measured for every call-path, as well as the respective execution time. Note that it is important to distinguish between inclusive and exclusive values of metrics. An *inclusive* metric value correponds to the execution of the function including all

---

[9] *CubePL* stands for **Cube P**rocessing **L**anguage.

function calls inside it (e.g., the inclusive metric value of *main* is the value measured for itself plus the values of *foo* and *bar*). An *exclusive* metric value only corresponds to the execution of the function itself excluding the values of functions called from it. In this example exclusive values are stored for every call-path.

To calculate the floating-point operations per second (FLOPS) as a derived metric for every call path, the naïve approach would be to define a new metric FLOPS and calculate its values for every call path using the formula $\frac{FLOP_c}{time_c}$, where c is either main, foo, or bar, and store the resulting values as a data metric within the corresponding Cube profile.

With this approach, however, a problem arises when the user would like to get an inclusive value of the metric FLOPS for the main call path. In this case, the Cube library would sum up all values of the metric FLOPS for every region and therefore deliver the result of the expression

$$\frac{FLOP_{main}}{time_{main}} + \frac{FLOP_{foo}}{time_{foo}} + \frac{FLOP_{bar}}{time_{bar}}$$

However, this expression is the *sum of FLOPS of every region* instead of the intended *FLOPS of main*. So, the correct calculation is

$$\frac{FLOP_{main} + FLOP_{foo} + FLOP_{bar}}{time_{main} + time_{foo} + time_{bar}}$$

which gives the desired floating-point operations per second for main.

As can be seen, the calculation of the $\frac{FLOP}{time}$ ratio should be performed as a last step – after the aggregation of the corresponding metric values FLOP and time is done. We call this a *postderived* metric. In contrast, the case where the derived metric should be calculated before the aggregation of the involved metrics is called a *prederived* metric.

The corresponding *CubePL* expression for the postderived FLOPS metric reads:

```
metric::flop()/metric::time()
```

As another slightly more complicated example of a postderived metric, consider the *CubePL* expression for the average time per visit of a function:

```
metric::time(i)/metric::visits(e)
```

Here the inclusive value of the time metric has to be used to include the time spent in child functions. However, for the visits metric, the exclusive value is necessary to only account for the calls to the function itself.

Generally, we advise users to start with the examples shipped with Cube and use those as a starting point to develop their own derived metrics.

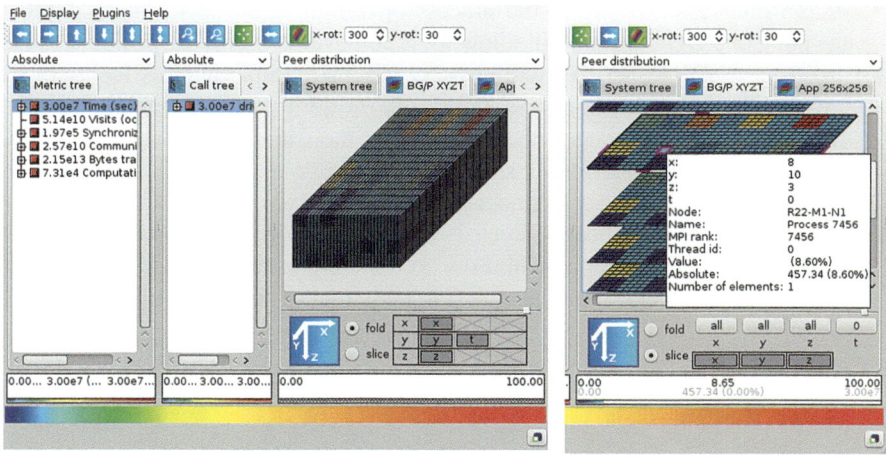

Fig. 2 Cube screenshot showing folding and slicing of multi-dimensional Cartesian topologies

4.3.2 Visualization of High-Dimensional Topologies in the Cube GUI

Many parallel applications define *virtual topologies* to express neighborhood relationships between processes, e.g., based on the chosen domain decomposition. Often, such virtual topologies are specified as multi-dimensional Cartesian grids. Another type of topologies are *physical topologies* reflecting the structure of the hardware the application run on. A typical three-dimensional physical topology is given by the (hardware) nodes in the first dimension, and the arrangement of cores/processors on nodes in further two dimensions. Further, some machines exhibit specific network layouts, like the IBM Blue Gene/Q 5-D torus network.

Until version 3, Cube was only able to handle Cartesian topologies up to three dimensions (where the folding of the Blue Gene tori into 3 dimensions was hard-coded). This situation was significantly improved in Cube v4. The Cube display now supports multi-dimensional Cartesian grids, where grids with higher dimensionality can be sliced or folded down to two or three dimensions for presentation – where the details of slicing and folding can be controlled by the user. An example of a visualization of a four-dimensional topology[10] is shown on Fig. 2.

4.3.3 Future Plans

Cube v4 is under continuous development, with two major feature additions coming up in the near future:

---

[10]Note that the topology toolbar is only enabled when a topology is available to be displayed.

- Plugin interface: We are in the process of designing an open plugin interface for the Cube GUI, which will allow users to develop problem-specific analyses based on CUBE4 data. We will provide various example plugins for users and are planning to establish an open repository for third-party plugins.
- Another major milestone is the transition of the Cube GUI from a monolithic architecture to a client/server architecture. Here, a client running on, e.g., the user's desktop machine can connect to a server started on the HPC system where the experiments have been collected. This will remove the necessity to transfer large Cube files between machines or to remotely execute the Cube GUI using X11 forwarding, and should also lead to increased scalability.

5   Conclusion

Like any other actively used software framework, the Scalasca parallel performance analysis toolset is constantly under development. Besides fixing bugs uncovered by our rigorous testing or reported by users, we also develop new features – often inspired by feedback from our active user community. In addition, the HPC landscape is also constantly changing due to the introduction of new computer systems and components, which in turn require new versions of parallel programming paradigms or sometimes even trigger the invention of new paradigms requiring further changes, additions, and adaptations in all performance tools. To better cope with this situation, the Scalasca v2 series is now based on the community-developed performance measurement and runtime infrastructure Score-P.

This change required significant redesign and implementation efforts of the Scalasca tool components as outlined in this paper. However, we believe that we were successful in minimizing the user-visible changes to the absolute necessary. After one year of deployment of the new Score-P based version of Scalasca, we see already a much quicker adaption of the infrastructure to new architectures and paradigms, with much more to come in the short term. Sharing development efforts for the parallel program instrumentation and measurement components with other tool developers also allowed us to provide better documentation as well as to improve the development and testing processes, while at the same time freeing some highly-needed resources for new and improved analysis features.

Acknowledgements The authors would like to use this opportunity to thank the HPC community in general and the users of our tools in particular for their feedback, feature requests, and bug reports. Without them, our tools would not be as good as they are now. We also want to thank the entire Scalasca and Score-P development teams for many fruitful discussions, insights, and sharing their experiences.

References

1. Benedict, S., Petkov, V., Gerndt, M.: PERISCOPE: an online-based distributed performance analysis tool. In: Müller, M.S., Resch, M.M., Schulz, A., Nagel, W.E. (eds.) Tools for High Performance Computing 2009, pp. 1–16. Springer, Berlin/Heidelberg (2010)
2. Böhme, D., Geimer, M., Wolf, F., Arnold, L.: Identifying the root causes of wait states in large-scale parallel applications. In: Proceedings of the 39th International Conference on Parallel Processing (ICPP), San Diego, pp. 90–100. IEEE Computer Society (2010)
3. Böhme, D., de Supinski, B.R., Geimer, M., Schulz, M., Wolf, F.: Scalable critical-path based performance analysis. In: Proceedings of the 26th IEEE International Parallel & Distributed Processing Symposium (IPDPS), Shanghai, pp. 1330–1340. IEEE Computer Society (2012)
4. Bueno, J., Planas, J., Duran, A., Badia, R., Martorell, X., Ayguade, E., Labarta, J.: Productive programming of GPU clusters with OmpSs. In: Proceedings of the 26th IEEE International Parallel Distributed Processing Symposium (IPDPS), Shanghai, pp. 557–568. IEEE Computer Society (2012)
5. CAPS, CRAY, NVIDIA, PGI: The OpenACC application programming interface. http://www.openacc.org/sites/default/files/OpenACC%202%200.pdf (2013)
6. Eichenberger, A.E., Mellor-Crummey, J.M., Schulz, M., Wong, M., Copty, N., DelSignore, J., Dietrich, R., Liu, X., Loh, E., Lorenz, D.: OMPT: OpenMP tools application programming interfaces for performance analysis. In: Proceedings of the 9th International Workshop on OpenMP (IWOMP), Canberra. LNCS, vol. 8122, pp. 171–185. Springer, Berlin/Heidelberg (2013)
7. Eschweiler, D., Wagner, M., Geimer, M., Knüpfer, A., Nagel, W.E., Wolf, F.: Open trace format 2 – the next generation of scalable trace formats and support libraries. In: Proceedings of the International Conference on Parallel Computing (ParCo), Ghent. Advances in Parallel Computing, vol. 22, pp. 481–490. IOS Press (2012)
8. Frings, W., Wolf, F., Petkov, V.: Scalable massively parallel I/O to task-local files. In: Proceedings of ACM/IEEE SC09 Conference, Portland (2009)
9. Geimer, M., Shende, S.S., Malony, A.D., Wolf, F.: A generic and configurable source-code instrumentation component. In: Allen, G., Nabrzyski, J., Seidel, E., van Albada, G.D., Dongarra, J., Sloot, P.M.A. (eds.) Proceedings of the International Conference on Computational Science (ICCS), Baton Rouge. Lecture Notes in Computer Science, vol. 5545, pp. 696–705. Springer (2009)
10. Geimer, M., Wolf, F., Wylie, B.J.N., Ábrahám, E., Becker, D., Mohr, B.: The *Scalasca* performance toolset architecture. Concurr. Comput.: Pract. Exp. **22**(6), 702–719 (2010)
11. Gleim, U., Levy, M.: MTAPI: parallel programming for embedded multicore systems. http://www.multicore-association.org/pdf/MTAPI_Overview_2013.pdf (2013)
12. Ilsche, T., et al.: Combining instrumentation and sampling for trace-based application performance analysis. In: Proceedings of 8th Parallel Tools Workshop, Stuttgart. Springer (To appear)
13. Jiang, J., Philippen, P., Knobloch, M., Mohr, B.: Performance measurement and analysis of transactional memory and speculative execution on IBM Blue Gene/Q. In: Proceedings of the 20th Euro-Par Conference, Porto. Lecture Notes in Computer Science, vol. 8632, pp. 26–37. Springer (2014)
14. Knüpfer, A., Brunst, H., Doleschal, J., Jurenz, M., Lieber, M., Mickler, H., Müller, M.S., Nagel, W.E.: The Vampir performance analysis toolset. In: Tools for High Performance Computing (Proceedings of the 2nd Parallel Tools Workshop, July 2008, Stuttgart), pp. 139–155. Springer (2008)
15. Knüpfer, A., Rössel, C., an Mey, D., Biersdorff, S., Diethelm, K., Eschweiler, D., Geimer, M., Gerndt, M., Lorenz, D., Malony, A.D., Nagel, W.E., Oleynik, Y., Philippen, P., Saviankou, P., Schmidl, D., Shende, S.S., Tschüter, R., Wagner, M., Wesarg, B., Wolf, F.: Score-P – a joint performance measurement run-time infrastructure for Periscope, Scalasca, TAU, and Vampir. In: Proceedings of 5th Parallel Tools Workshop, Dresden, pp. 79–91. Springer (2012)

16. Luk, C.K., Cohn, R., Muth, R., Patil, H., Klauser, A., Lowney, G., Wallace, S., Reddi, V.J., Hazelwood, K.: Pin: building customized program analysis tools with dynamic instrumentation. SIGPLAN Not. **40**(6), 190–200 (2005)
17. Massage Passing Interface Forum: MPI: a message-passing interface standard version 3.0. http://www.mpi-forum.org/docs/mpi-3.0/mpi30-report.pdf (2012)
18. Mohr, B., Malony, A.D., Hoppe, H.C., Schlimbach, F., Haab, G., Hoeflinger, J., Shah, S.: A performance monitoring interface for OpenMP. In: Proceedings of Fourth European Workshop on OpenMP (EWOMP), Rome (2002)
19. NVIDIA: CUDA toolkit documentation. http://docs.nvidia.com/cuda/ (2014)
20. OPARI2 web page. http://www.vi-hps.org/tools/opari2.html (2014)
21. OpenMP Architecture Review Board: The OpenMP API specification for parallel programming, version 4.0. http://openmp.org/wp/openmp-specifications/ (2013)
22. PerfExplorer web page. http://www.cs.uoregon.edu/research/tau/docs/perfexplorer/ (2014)
23. Scalasca Performance Analysis Toolset web page. http://www.scalasca.org (2014)
24. Score-P User Manual. Available as part of the Score-P installation or online at http://www.score-p.org (2014)
25. Shende, S., Malony, A.D.: The TAU parallel performance system. Int. J. High Perform. Comput. Appl. **20**(2), 287–311 (2006)
26. The Khronos Group: OpenCL 2.0 API specification. https://www.khronos.org/registry/cl/specs/opencl-2.0.pdf (2014)
27. Vampir web page. http://www.vampir.eu/ (2014)
28. Zhukov, I., Wylie, B.J.N.: Assessing measurement and analysis performance and scalability of Scalasca 2.0. In: Proceedings of the Euro-Par 2013: Parallel Processing Workshops, Aachen. LNCS, vol. 8374, pp. 627–636. Springer (2014)

Allinea MAP: Adding Energy and OpenMP Profiling Without Increasing Overhead

Christopher January, Jonathan Byrd, Xavier Oró, and Mark O'Connor

Abstract Allinea MAP was introduced in 2013 as a highly scalable, commercially-supported sampling-based MPI profiler that tracks performance data over time and relates it directly to the program source code. We have since extended its capabilities to support profiling of OpenMP regions and POSIX threads (pthreads) in general. We will show the principles we used to highlight the balance between multi-core (OpenMP) computation, MPI communication and serial code in Allinea MAP's updated GUI. Graphs detailing performance metrics (memory, IO, vectorised operations etc.) complete the performance profile.

We have also added power-usage metrics to Allinea MAP and are actively seeking collaboration with vendors, application users and other tools writers to define how best HPC can meet the power requirements moving towards exascale. MAP's data is provided for export to other tools and analysis in an open XML-based format.

1 Introduction

Allinea MAP is a software performance profiling tool introduced by Allinea in 2013 that provides an intuitive, easy to use overview of the performance of HPC programs. MAP focuses on low impact measurement with aggregation across processes resulting in low run-time overhead and small output files. MAP is built on the same petascale-capable scalable infrastructure [1] as Allinea DDT.

OpenMP [2] is a directive-based multi-language standard for high-level parallelism, including shared memory programming. Many HPC programs use hybrid MPI and OpenMPI parallel programming in which MPI is used for communication between compute nodes and shared memory programming is used inside each SMP node.

C. January (✉) • J. Byrd • X. Oró
Allinea Software Ltd., Warwick, UK
e-mail: chris.january@allinea.com

M. O'Connor
Allinea Software GmbH, München, Germany

© Springer International Publishing Switzerland 2015

C. Niethammer et al. (eds.), *Tools for High Performance Computing 2014*,
DOI 10.1007/978-3-319-16012-2_2

25

2   Multi-threading and OpenMP in MAP

Allinea MAP has scalable support for MPI programs, aggregating data across processes so the size of the output file is asymptotic with respect to the number of processes. We have extended its capabilities to support profiling of OpenMP regions and POSIX threads in general.

2.1   Application Activity

Reducing the amount of computation required to achieve a given result inevitably results in shorter run-times for a job. Shorter run-times, in turn, allow faster turnaround and reduced cost. To this end Allinea MAP shows, for example, a breakdown of where computation occurs by function, and even source line, so that a user can spot the best candidates for optimization to reduce compute time.

That having been said, one of the primary principles behind the design of Allinea MAP is that users want to maximise the use of the available compute resources, that is to say, maximise the amount of time an HPC system is actually computing. Idle CPU cores are core hours wasted.

The Allinea MAP interface displays *time glyphs* that show how a program's activity varies over time. MAP graphs all of a program's activity against the same common axis—wall time. When profiling an MPI program the time glyphs show how much time is spent in MPI calls vs computation. This allows a user to see hot spots in their program where MPI communication is stalling, for example. We have extended this concept to MAP's profiling of OpenMP regions and POSIX threads— MAP shows the time split between multi-core, MPI and serial code.

The *Application Activity* view shows the overall activity of the program over time—across all processes and the entire source code. The view shows how the distribution of time in multi-core, MPI and serial code changes as the program progresses, and each region of time can be tied to the lines of source code that were executing during it.

Figure 5 shows the full Allinea MAP user interface with the Application Activity view at the top of the window. The program's activity is further broken down by source line. MAP shows the program's source code front and centre and against each source line is a time glyph of the program's activity on that line, and the functions called by that line.

For performance reasons many MPI and OpenMP instrumentations busy wait when waiting for messages or work, rather than sleeping. MAP treats all time spent in an MPI function as MPI activity, regardless of whether a process is running or sleeping. Similarly MAP treats all time spent in the OpenMP runtime as 'OpenMP Overhead'. MAP distinguishes between calls to the OpenMP runtime from inside an active OpenMP work share and those outside. Calls from inside a parallel region,

Fig. 1 Application Activity view for an inefficient program. Key: *Green* OpenMP, *Grey* OpenMP Overhead, *Blue* MPI

Fig. 2 Application Activity view for an optimized version of the same program. Key: *Green* OpenMP, *Grey* OpenMP Overhead, *Blue* MPI

for example, are associated with the region in question and the time spent in those calls is shown as 'OpenMP Overhead' against the entry point to the parallel region.

A program that uses OpenMP will rarely spend all its runtime in parallel code. Instead it may have a mixture of serial regions, running on the main thread only, and parallel regions running on more than one thread. MAP distinguishes between these serial and parallel divisions. When the program is in a parallel region MAP gives equal weight to each thread. Inside a serial region time, however, spent in the OpenMP runtime by other threads is ignored and the main thread is given full weight.

Figure 1 shows the Application Activity view for a hybrid MPI/OpenMP program. The *x* axis corresponds to wall clock time and the *y* axis to the number of cores in OpenMP regions. The areas marked in grey show cores busy waiting in the OpenMP runtime. Much of the available compute power is being left on the table.

Figure 2 shows the Application Activity view for an optimized version of the same program. In this example the available compute resources are much better utilised.

2.2 OpenMP Regions

Allinea MAP uses lightweight instrumentation of MPI functions to, for example, accurately record the time spent in MPI function calls. In order to support profiling of OpenMP programs we have added lightweight instrumentation of some OpenMP runtime functions. Unlike source to source instrumentors such as Opari [7], MAP uses binary function interception. In many cases a program does not need to be recompiled at all for use with MAP; in the worst case it needs to be relinked.

The OpenMP regions and other work sharing constructs are shown in a new view, the *OpenMP Regions* view. Most work sharing constructs are supported, with the notable exception of OpenMP tasks.

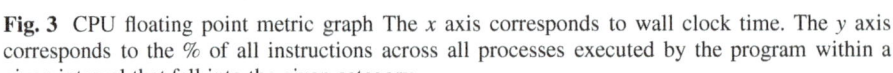

Fig. 3 CPU floating point metric graph The *x* axis corresponds to wall clock time. The *y* axis corresponds to the % of all instructions across all processes executed by the program within a given interval that fell into the given category

The OpenMP runtime instrumentation also allows MAP to associate code running on an OpenMP worker thread with an invoking call site on the main thread (or other worker thread in the case of nested parallelism). Code running on the worker threads is reparented by the GUI so it appears under the invoking call site in the *Stacks* view.

2.3 CPU Metrics

Allinea MAP samples the class of CPU instructions being executed by a program and displays these as *CPU metric graphs* in the MAP user interface. The graphs show the fraction of compute resources spent in a particular class of CPU instruction (e.g. floating point as shown in Fig. 3).

A surprisingly useful metric is the fraction of compute resources spent in floating point vector operations. The default compiler options for many compilers result in less vectorisation than may be expected. MAP can not only expose this, but also be used to measure the effect of different compiler options on performance.

We have extended MAP's CPU instruction sampling to multi-core programs. Following the principle that users want to maximise the use of *available* compute resources, MAP effectively samples *per core* rather than per thread. Cores that are idle are marked as such so that the MAP CPU metrics show the fraction of *available compute resources* spent in a particular class of CPU instruction.

The available compute resources are initially assumed to be *min(logical cores, max(active cores))*, although the user may override this value in the user interface.

2.4 Worked Example

Figure 4 shows how the performance of the miniMD hybrid MPI/OpenMP benchmark code varies with the number of OpenMP threads on a Cray XK7.

miniMD contains two versions of the code—one OpenMP, and one serial—that are selected at runtime. The same executable is used in both cases. As can be seen the hybrid MPI/OpenMP kernel performs considerably worse than the pure MPI kernel. Using Allinea MAP we may explore why this is the case.

Figure 5 shows an Allinea MAP profile of a miniMD hybrid MPI/OpenMP run with one MPI process per node and 16 threads per MPI process.

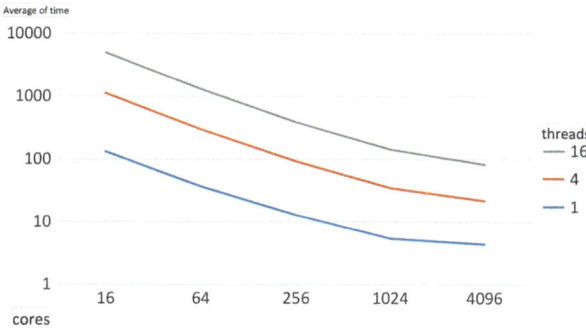

Fig. 4 miniMD Performance on a Cray XK7 (16 cores per compute node; 10,000 steps). The processes per node was adjusted according to the number of OpenMP threads so the number of cores in use on each node remained constant

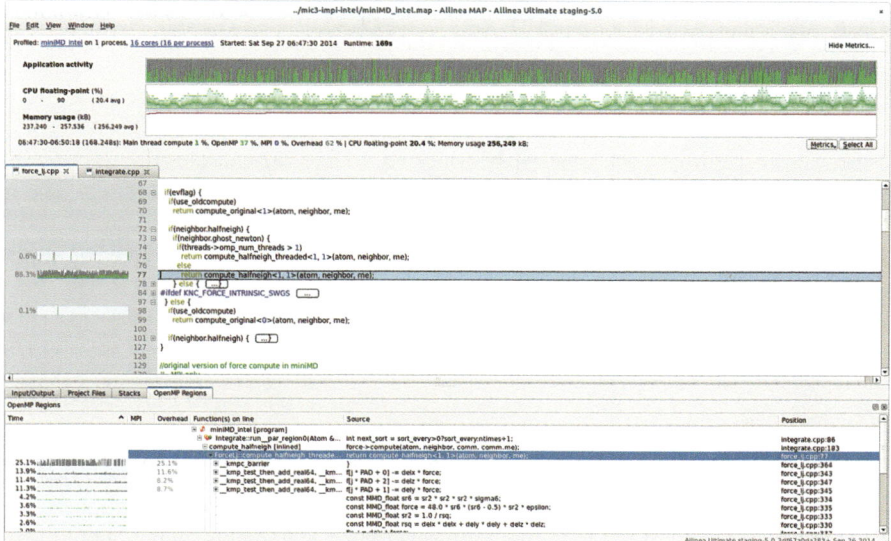

Fig. 5 miniMD profile in Allinea MAP

At the top of the window is the Application Activity view. Figure 6 shows the view in more detail. The area shown in green represents OpenMP compute time. The majority of the activity shows in grey—this represents OpenMP overhead, i.e. time spent in the OpenMP runtime, and as can be seen this represents 62 % of the program's run time.

Figure 7 uses the OpenMP Regions view to drill down to see how the OpenMP region is spending its time. Most of the time is spent in just four lines. A lot of time is spent in the barrier at the end of the parallel for loop. This may indicate the OpenMP threads are unbalanced and some are doing more work than others.

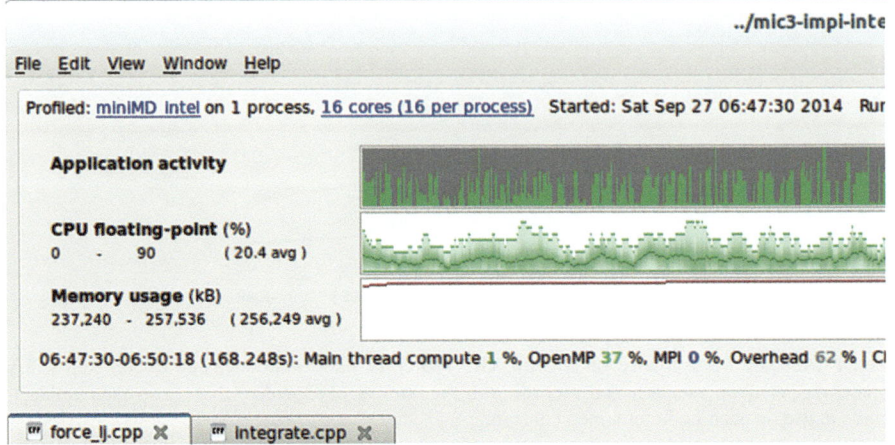

Fig. 6 Application Activity for miniMD. Key: *Green* OpenMP, *Grey* OpenMP Overhead, *Blue* MPI

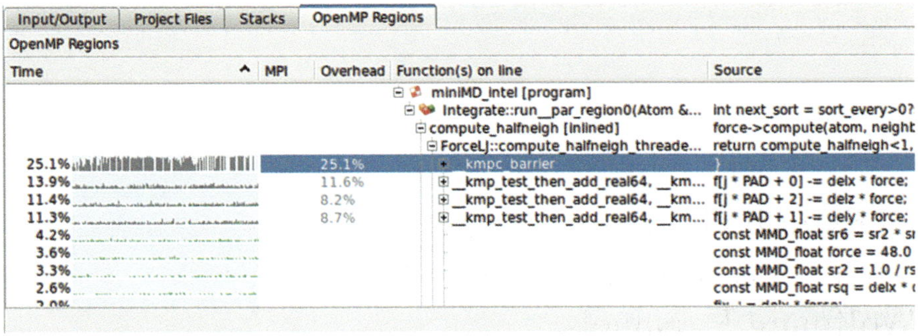

Fig. 7 OpenMP Regions view showing most of the program's runtime is spent in just four lines

Looking at the body of the loop there are several `if` statements, which would lead to unbalanced workloads, that support this idea.

As can be seen in Fig. 8 a lot of time is also spent on the three lines that are preceded by `#pragma omp atomic`. In this case the `atomic` directive has two side effects:

1. Effectively serialises the code because the atomic operations require a memory barrier.
2. Prevents vectorization.

MAP's CPU instructions view in Fig. 9 shows that no vectorised floating point instructions are used in the program. This may be contrasted to the pure MPI version in Fig. 10. The compiler can vectorize the latter because it does not use OpenMP atomics.

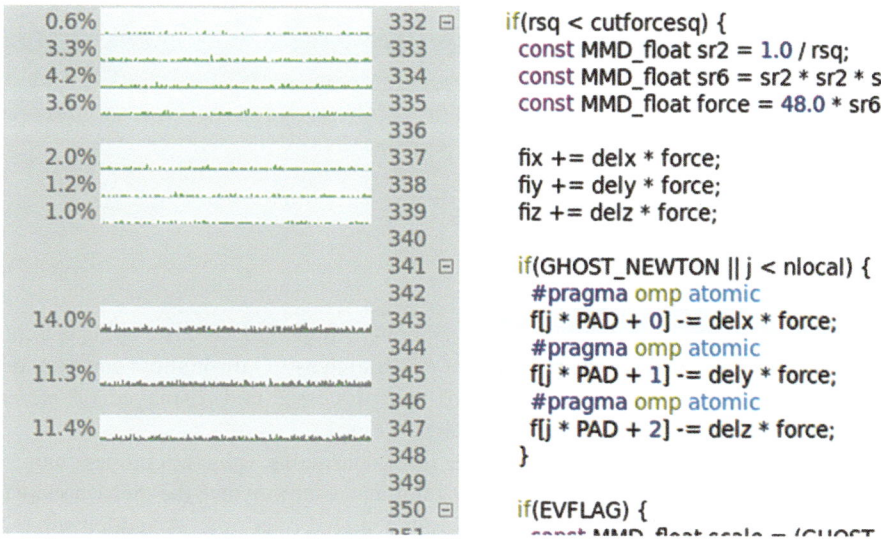

0.6%	332 ⊟	if(rsq < cutforcesq) {
3.3%	333	const MMD_float sr2 = 1.0 / rsq;
4.2%	334	const MMD_float sr6 = sr2 * sr2 * s
3.6%	335	const MMD_float force = 48.0 * sr6
	336	
2.0%	337	fix += delx * force;
1.2%	338	fiy += dely * force;
1.0%	339	fiz += delz * force;
	340	
	341 ⊟	if(GHOST_NEWTON || j < nlocal) {
	342	#pragma omp atomic
14.0%	343	f[j * PAD + 0] -= delx * force;
	344	#pragma omp atomic
11.3%	345	f[j * PAD + 1] -= dely * force;
	346	#pragma omp atomic
11.4%	347	f[j * PAD + 2] -= delz * force;
	348	}
	349	
	350 ⊟	if(EVFLAG) {
	351	const MMD float scale = (GHOST

Fig. 8 36.7 % of the program's runtime is spent on the three lines preceded by #pragma omp atomic

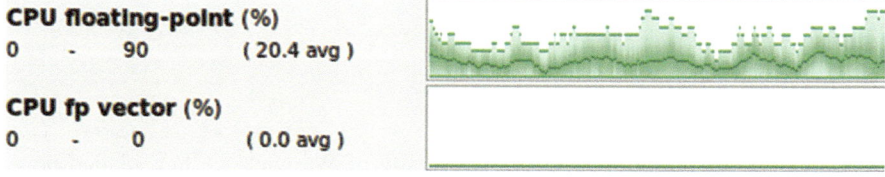

CPU floating-point (%)
0 - 90 (20.4 avg)

CPU fp vector (%)
0 - 0 (0.0 avg)

Fig. 9 0 % CPU floating point vector instructions used by the hybrid MPI/OpenMP kernel

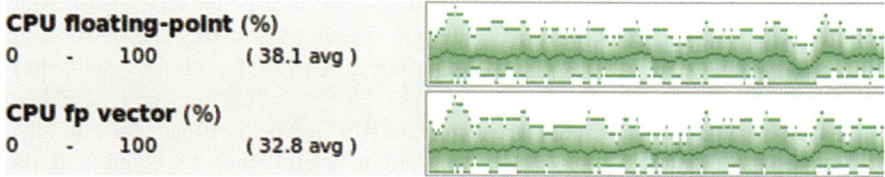

CPU floating-point (%)
0 - 100 (38.1 avg)

CPU fp vector (%)
0 - 100 (32.8 avg)

Fig. 10 32.8 % mean CPU floating point vector instructions used by the pure MPI kernel

In conclusion Allinea MAP allows us to determine the reasons why the hybrid MPI + OpenMP version of the miniMD kernel performs much more poorly than the pure MPI one:

- Use of OpenMP atomics.
- Lack of vectorization.
- Threads not balanced.

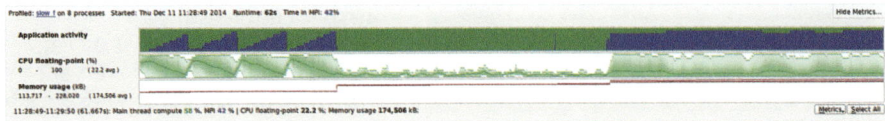

Fig. 11 Metric graphs in Allinea MAP

2.5 Performance Metrics

In addition to the information detailed above Allinea MAP also collects other performance metrics over the run-time of a job, such as I/O throughput and memory usage. The values are aggregated across all processes and displayed on *metric graphs* in the MAP user interface as shown in Fig. 11.

The metric graphs allows a user to see how a particular value or counter changed over the run-time of their program. For example, a user may use the metrics view to see how the memory usage of their program changed over time. A sudden spike in memory usage may indicate a problem that needs investigating.

We have extended Allinea MAP to collect a new class of measurement: *energy metrics*.

3 Energy Metrics

As part of a collaboration with the University of Warwick we have added power usage metric graphs to MAP that show the power usage of a program over time. The principle source of data is the Intel RAPL [3] energy counters included in recent Intel CPUs. We have also added support for the Intel Energy Checker SDK [5] which includes a number of drivers for external power meters, allowing the power usage of not only the CPU, but a whole compute node, or even the whole system, to be recorded.

One way of reducing the energy usage of an HPC job is to offload some, or all, of the computation to accelerator cards which are more efficient for a given workload. In addition to the Intel RAPL energy counters for regular Intel CPUs we have also added power metrics for Intel Xeon Phi accelerator cards and NVIDIA CUDA cards using the respective system management interfaces.

How best to use power usage metrics and other performance metrics to reduce the energy consumption of a job is an open question. The biggest factor influencing the power consumption of a general purpose CPU is the frequency it is running at. Dynamic frequency scaling is a possible route to reduced energy consumption [6]. For example: many MPI implementations busy-wait for MPI communications to complete. The frequency of the CPU can be turned down at these times. MAP provides the tools to not only identify when a program is waiting for MPI communication to finish (and how often, and how much), but also to measure the impact on both total energy consumption and runtime if frequency scaling is applied at those times.

3.1 Future Work

CPU and accelerator cards are just part of the story. There are many other sources of power consumption, both within individual compute nodes, and at other points in an HPC system. A number of system vendors either already provide their own power management counters, or are planning on doing so in the future (e.g. Cray's Power Management Counters [8]). We are looking to add support for these and to partner with 3rd parties who have developed, or are thinking of developing, frameworks for measuring power consumption.

4 Custom Metrics

To make it easy for third parties to add metrics to Allinea MAP we have developed support for *metric plugins* which provide *custom metrics* that are recorded in addition to MAP's built-in metrics. This allows a user to see information about values and counters that are not recorded by MAP out of the box, even values specific to a particular program.

A metric plugin consists of:

- An XML file that describes the custom metrics that the metric plugin provides
- A shared library that contains the code to sample the custom metrics

The XML descriptor arranges for a function—sample_interrupts in the example below—in the shared library to be called every time MAP samples the user's program. This simple mechanism makes it very easy to add new metrics to MAP, or to integrate MAP with data sources from other tools. Custom metrics are pre process, rather than per thread, so the function is always called from the main thread.

Since the user's program will have been interrupted by the MAP sampler at an arbitrary point it is important for custom metrics functions to only call async-signal-safe functions [4] and thread safe functions. Many commonly used C library functions, such as malloc, are not async-signal-safe so MAP provides its own async-signal-safe replacements (e.g. allinea_safe_malloc).

4.1   Example: System Interrupt Count

The listing below shows the XML descriptor for a plugin that samples the number
of system interrupts taken.

```
<!-- version is the file format version -->
<metricdefinitions version="1">
  <!-- id is the internal name for this metric, as used in the .map XML -->
  <metric id="interrupts">
    <!-- Data type used to store the sample values Valid data types:
         - uint64_t
         - double -->
    <dataType>uint64_t</dataType>
    <!-- The domain the metric is to be sampled in.
         Only time is supported. -->
    <domain>time</domain>
    <!-- Sample source
         Specifies the source of data for this metric,
         i.e. a function in a shared library.

         The function signature depends on the dataType:
         - uint64_t:  int function(metric_id_t metricId,
                                    struct timespec inCurrentSampleTime,
                                    uint64_t *outValue);
         - double:    int function(metric_id_t metricId,
                                    struct timespec inCurrentSampleTime,
                                    double *outValue);

         If the result is undefined for some reason the function may
         return the special sentinel value ~0 (unsigned integers) or
         Nan (floating point)

         Return value is 0 if success, -1 if failure (and set errno)

         If divideBySampleTime is true then the values returned by outValue
         will be divided by the sample interval to get the final value. -->
    <source ref="sample_src"
            functionName="sample_interrupts"
            divideBySampleTime="true"/>

    <!-- Display attributes used by the GUI. -->
    <display>
      <!-- Display name for the metric as used in the GUI. -->
      <displayName>Interrupts</displayName>
      <!-- Brief description of the metric.. -->
      <description>Total number of system interrupts taken</description>
      <!-- The type of metric.
           This is used by the GUI to group metrics. -->
      <type>interrupts</type>
      <!-- The colour to use for the metric graphs for this metric. -->
      <colour>green</colour>
    </display>
  </metric>

    <!-- Metric group for interrupt metrics.,
         These groupings are used in the GUI. -->
    <metricGroup id="sample">
      <!-- Display name for the group as use din the GUI. -->
      <displayName>Sample</displayName>
      <!-- Brief description of the group. -->
      <description>Interrupt metrics</description>
      <!-- References to all the metrics included in the group. -->
      <metric ref="interrupts"/>
    </metricGroup>

    <!-- Definition of the sample source (metric plugin) used for the sample
         custom metric. -->
    <source id="sample_src">
      <!-- File name of the sample metric plugin shared library. -->
      <sharedLibrary>libsample.so</sharedLibrary>
    </source>
</metricdefinitions>
```

5 Summary

We have extended Allinea MAP to add profiling of OpenMP and POSIX threads by building on our original principles for MPI profiling. MAP is designed to help a user to maximise the use of available compute resources.

Energy consumption will continue to be an important consideration in HPC. MAP helps users not only optimize for minimal run-time, but also for minimal energy consumption.

The addition of custom metrics to MAP not only makes it easy for third parties to add their own metrics, but also makes it possible to expose program-specific values in the MAP user interface.

Users benefit when their tools are well integrated. Allinea is committed to continuing to open up MAP for integration with other tools.

References

1. Christopher January David Lecomber, M.O.: Debugging at petascale and beyond. In: Proceedings of Cray User Group 2011. Fairbanks, Alaska, USA (2011)
2. Dagum, L., Menon, R.: Openmp: an industry-standard api for shared-memory programming. IEEE Comput. Sci. Eng. **5**(1), 46–55. New York, NY, USA (1998). doi:10.1109/99.660313. http://dx.doi.org/10.1109/99.660313
3. David, H., Gorbatov, E., Hanebutte, U.R., Khanna, R., Le, C.: Rapl: memory power estimation and capping. In: Proceedings of the 16th ACM/IEEE International Symposium on Low Power Electronics and Design (ISLPED '10), pp. 189–194. ACM, New York (2010). doi:10.1145/1840845.1840883. http://doi.acm.org/10.1145/1840845.1840883
4. IEEE: IEEE Std 1003.1-2013 Standard for Information Technology—Portable Operating System Interface (POSIX) Base Specifications, Issue 7. IEEE (2013). http://pubs.opengroup.org/onlinepubs/9699919799/functions/V2_chap02.html
5. Intel energy checker sdk. http://software.intel.com/enus/articles/intel-energy-checker-sdk
6. Laros III, J.H., Pedretti, K.T., Kelly, S.M., Shu, W., Vaughan, C.T.: Energy based performance tuning for large scale high performance computing systems. In: Proceedings of the 2012 Symposium on High Performance Computing (HPC '12), pp. 6:1–6:10. Society for Computer Simulation International, San Diego, California, USA (2012). http://dl.acm.org/citation.cfm?id=2338816.2338822
7. Mohr, B., Malony, A.D., Shende, S., Wolf, F.: Design and prototype of a performance tool interface for openmp. J. Supercomput. **23**(1), 105–128 (2002). doi:10.1023/A:1015741304337. http://dx.doi.org/10.1023/A:1015741304337
8. Using Cray Performance Measurement and Analysis Tools, chap. 3.2.6.4 Power Management Counters. S2376613, Cray (2013)

DiscoPoP: A Profiling Tool to Identify Parallelization Opportunities

Zhen Li, Rohit Atre, Zia Ul-Huda, Ali Jannesari, and Felix Wolf

Abstract The stagnation of single-core performance leaves application developers with software parallelism as the only option to further benefit from Moore's Law. However, in view of the complexity of writing parallel programs, the parallelization of myriads of sequential legacy programs presents a serious economic challenge. A key task in this process is the identification of suitable parallelization targets in the source code. We have developed a tool called DiscoPoP showing how dependency profiling can be used to automatically identify potential parallelism in sequential programs. Our method is based on the notion of computational units, which are small sections of code following a read-compute-write pattern that can form the atoms of concurrent scheduling. DiscoPoP covers both loop and task parallelism. Experimental results show that reasonable speedups can be achieved by parallelizing sequential programs manually according to our findings. By comparing our findings to known parallel implementations of sequential programs, we demonstrate that we are able to detect the most important code locations to be parallelized.

1 Introduction

Although the component density of microprocessors is still rising according to Moores Law, single-core performance is stagnating for more than ten years now. As a consequence, extra transistors are invested into the replication of cores, resulting in the multi- and many-core architectures popular today. The only way for developers to take advantage of this trend if they want to speed up an individual application is to match the replicated hardware with thread-level parallelism. This, however, is often challenging especially if the sequential version was written by someone

Z. Li (✉) · Z. Ul-Huda · A. Jannesari · F. Wolf
German Research School for Simulation Sciences, 52062 Aachen, Germany
Technische Universität, Darmstadt, 64289 Darmstadt, Germany
e-mail: {li, ul-huda, jannesari, wolf}@cs.tu-darmstadt.de

R. Atre
Aachen Institute for Advanced Study in Computational Engineering Science, Aachen, Germany

RWTH Aachen University, Aachen, Germany
e-mail: atre@aices.rwth-aachen.de

© Springer International Publishing Switzerland 2015
C. Niethammer et al. (eds.), *Tools for High Performance Computing 2014*,
DOI 10.1007/978-3-319-16012-2_3

else. Unfortunately, in many organizations the latter is more the rule than the exception [1]. To find an entry point for the parallelization of an organization's application portfolio and lower the barrier to sustainable performance improvement, tools are needed that identify the most promising parallelization targets in the source code. These would not only reduce the required manual effort but also provide a psychological incentive for developers to get started and a structure for managers along which they can orchestrate parallelization workflows.

In this paper, we present an approach for the discovery of potential parallelism in sequential programs that—to the best of our knowledge—is the first one to combine the following elements in a single tool:

1. Detection of available parallelism with high accuracy
2. Identification of code sections that can run in parallel, supporting the definition of parallel tasks—even if they are scattered across the code
3. Ranking of parallelization opportunities to draw attention to the most promising parallelization targets
4. Time and memory overhead that is low enough to deal with input programs of realistic size

Our tool, which we call DiscoPoP (= Discovery of Potential Parallelism), reverses the idea of data-race detectors. It profiles dependencies, but instead of only reporting their violation it also watches out for their absence. We use the dependency information to represent program execution as a graph, from which parallelization opportunities can be easily derived or based on which their absence can be explained. Since we track dependencies across the entire program execution, we can find parallel tasks even if they are widely distributed across the program or not properly embedded in language constructs, fulfilling the second requirement. To meet the third requirement, our ranking method considers a combination of execution-time coverage, critical-path length, and available concurrency. Together, these four properties bring our approach closer to what a user needs than alternative methods [2–4] do. We expand on earlier work [5], which introduced the algorithm for building the dependency graph based on the notion of computational units— which was, at the time a purely dynamic approach with significant time and memory overhead. That is why this paper concentrates mainly on the overall workflow, including a preceding static analysis, the minimization of runtime overhead, the ranking algorithm, and an evaluation using realistic examples along with a demonstration of identifying non-obvious tasks.

The remainder of the paper is structured as follows: In the next section, we review related work and highlight the most important differences to our own. In Sect. 3, we explain our approach in more detail. In the evaluation in Sect. 4, we run the NAS parallel benchmarks [6], a collection of programs derived from real CFD codes, to analyze the accuracy at which we identify and rank parallelism in spite of the optimizations we apply. Also, we show how we find parallel tasks and pipeline patterns that are not trivial to spot. Finally, we quantify the overhead of our tool both in terms of time and memory. Section 5 summarizes our results and discusses further improvements.

2 Related Work

After purely static approaches including auto-parallelizing compilers had turned out to be too conservative for the parallelization of general-purpose programs, a range of predominantly dynamic approaches emerged. Such dynamic approaches can be broadly divided into two categories. Tools in the first merely count dependencies, whereas tools in the second, including our own, exploit explicit dependency information to provide detailed feedback on parallelization opportunities or obstacles.

Kremlin [4] belongs to the first category. Using dependency information, it determines the length of the critical path in a given code region. Based on this knowledge, it calculates a metric called self-parallelism, which quantifies the parallelism of a code region. Kremlin ranks code regions according to this metric. Alchemist [3] follows a similar strategy. Built on top of Valgrind, it calculates the number of instructions and the number of violating read-after-write (RAW) dependencies across all program constructs. If the number of instructions of a construct is high while the number of RAW dependencies is low, it is considered to be a good candidate for parallelization. In comparison to our own approach, both Kremlin and Alchemist have two major disadvantages: First, they discover parallelism only at the level of language constructs, that is, between two predefined points in the code, potentially ignoring parallel tasks not well aligned with the source-code structure. Second, they merely quantify parallelism but do neither identify the tasks to run in parallel unless it is trivial as in loops nor do they point out parallelization obstacles.

Like DiscoPoP, Parwiz [2] belongs to the second category. It records data dependencies and attaches them to the nodes of the execution tree (i.e., a generalized call tree that also includes basic blocks) it maintains. In comparison to DiscoPoP, Parwiz lacks a ranking mechanism and does not explicitly identify tasks. They have to be manually derived from the dependency graph, which is demonstrated using small text-book examples.

Reducing the significant space overhead of tracing memory accesses was also successfully pursued in SD3 [7]. An essential idea that arose from there is the dynamic compression of strided accesses using a finite state machine. Obviously, this approach trades time for space. In contrast to SD3, DiscoPoP leverages an acceptable approximation, sacrificing a negligible amount of accuracy instead of time. The work from Moseley et al. [8] is a representative example of this approach. Sampling also falls into this category.

Prospector [9] is a parallelism-discovery tool based on SD3. It tells whether a loop can be parallelized, and provides a detailed dependency analysis of the loop body. It also tries to find pipeline parallelism in loops. However, no evaluation result or example is given for this feature.

3   Approach

Figure 1 shows the work flow of DiscoPoP. The work flow of DiscoPoP is divided into two phases: In the first phase, we instrument the target program and execute it. Control flow information and data dependencies are obtained in this phase. In the second phase, we build computational units (CUs) for the target program, and search for potential parallelism based on the CUs and dependence among them. The output is a list of parallelization opportunities, consisting of several code sections that may run in parallel. These opportunities are also ranked to allow the users focus on the most interesting opportunities.

3.1   Dependence Profiling

Data dependences can be obtained in two major ways: static and dynamic analysis. Static approaches determine data dependences without executing the program. Although they are fast and even allow fully automatic parallelization in some restricted cases [10, 11], they lack the ability to track dynamically allocated memory, pointers, and dynamically calculated array indices, which usually makes their assessment too pessimistic for practical purposes. In contrast, dynamic dependence profiling captures only those dependences that actually occur at runtime. Although dependence profiling is inherently input sensitive, the results are still useful in many situations, which is why such profiling forms the basis of many program analysis tools [2, 4, 9]. Besides, input sensitivity can be addressed by running the target program with changing inputs and computing the union of all collected dependences.

Dependence profiling component serves as the foundation of our tool. The profiler produces the following information:

- Pair-wise data dependencies
- Source code locations of dependencies and the names of the variables involved
- Runtime control-flow information

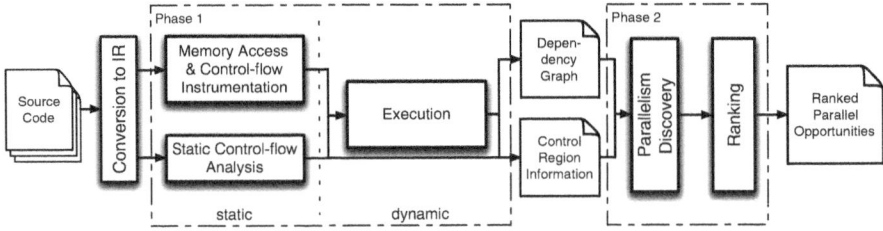

Fig. 1 The work flow of DiscoPoP

```
1:60 BGN loop
1:60 NOM  {RAW 1:vim}  {WAR 1:60|i}  {INIT *}
1:63 NOM  {RAW 1:59|temp1}  {RAW 1:67|temp1}
1:64 NOM  {RAW 1:60|i}
1:65 NOM  {RAW 1:59|temp1}  {RAW 1:67|temp1}  {WAR 1:67|temp2}  {INIT *}
1:66 NOM  {RAW 1:59|temp1}  {RAW 1:65|temp2}  {RAW 1:67|temp1}  {INIT *}
1:67 NOM  {RAW 1:65|temp2}  {WAR 1:66|temp1}
1:70 NOM  {RAW 1:67|temp1}  {INIT *}
1:74 NOM  {RAW 1:41|block}
1:74 END loop 1200
```

Fig. 2 A fragment of profiled data dependencies in a sequential program

We profile detailed pair-wise data dependencies because we do not want to lose the chance to report root causes that preventing parallelism. If detailed information is not required by a certain analysis, dependencies can be easily merged into coarser grain with the help of control-flow and variable name information, for example, dependencies between loops, between functions, or between objects. Control-flow information is necessary for building computational units.

A fragment of profiled data is shown in Fig. 2. A data dependency is represented as a triple `<sink, type, source>`. `type` is the dependency type (RAW, WAR or WAW). Note that a special type `INIT` represents the first write operation to a memory address. In this case, `source` of the dependency is empty, which is represented as `*`.

`sink` and `source` are the source code locations of the latter and the former memory accesses, respectively. `sink` is further represented as a pair `<fileID:lineID>`, while `source` is represented as a triple `<fileID:lineID|variableName>`. As it is shown in Fig. 2, data dependences with the same `sink` are aggregated together.

Identifier `NOM` (short for "NORMAL") means that the source line specified by aggregated `sink` has no control-flow information. Otherwise, `BGN` and `END` represent the entry and exit point of a control region, respectively. In Fig. 2, a loop starts at source line 1:60 and ends at source line 1:74. The number following `END loop` shows the actual number of iterations executed, which is 1,200 in this case.

In order to get pair-wise data dependencies dynamically, every load and store instruction is instrumented. Entry and exit points of control regions are determined statically, but loops are still instrumented so that the number of executed iterations can be recorded. Source code location and variable names are obtained with the help of debug symbols, thus the compiler option `-g` must be specified to compile the program.

3.2 Computational Unit

During the second phase, we search for potential parallelism based on the output of the first phase, which is essentially a graph of dependencies between source lines.

This graph is then transformed into another graph, whose nodes are parts of the code without parallelism-preventing read-after-write (RAW) dependencies inside. We call these nodes *computational units* (CUs). Based on this CU graph, we can detect potential parallelism and already identify tasks that can run in parallel.

A CU is defined as a set of instructions that form a *read-compute-write* pattern. A CU differs from the basic block such that a basic block contains operations that are consecutive and has only one entry and one exit point. A CU however, is a group of instructions that are not necessarily consecutive but perform a computation and is based on the use of a set of variables. A single CU or a group of CUs merged together can provide the code sections that perform a task. These code sections can be examined to see if they can be run concurrently with other code sections or themselves to exploit the available parallelism.

To understand what a CU is, consider the example in Fig. 3. The source lines 4 and 8 perform the initialization of variables *id_a* and *id_b* with a random value. Lines 5 and 9 perform the task of calculating *elem_a* and *elem_b* by using the value of *id_a* and *id_b* respectively. These two operations are performed independent of one another. The *while* loop in the function is responsible for checking if the two random values are equal and reassigning *id_b* if it is true. Finally, the lines 16 and 17 are responsible for writing the final computation back to **a* and **b*. In essence, the group of LLVM-IR instructions corresponding to the lines {4, 5, 16} perform one computation and the ones corresponding to the lines {8, 9, 13, 14, 17} perform another computation and these tasks are independent of each other except for the equality check condition of the *while* loop. In case of the *while* loop, identifying code sections that require synchronization or replication across concurrent threads will part of our future work.

The two computations mentioned above follow a basic rule where a variable or a group of variables are read and then they are used to perform another calculation. This is followed by the final state being written to another variable as a store operation. Hence, these two computations can be said to follow a *read-compute-write* pattern. The two computations can be visualized as seen in the Fig. 4. The final

```
1    void netlist:: (netlist_elem** a, netlist_elem** b, Rng* rng)
2    {
3        //get a random element
4        long id_a = rng->rand(_chip_size);
5        netlist_elem* elem_a = &(_elements[id_a]);
6
7        //now do the same for b
8        long id_b = rng->rand(_chip_size);
9        netlist_elem* elem_b = &(_elements[id_b]);
10
11       while (id_b == id_a)
12       {
13           id_b = rng->rand(_chip_size);
14           elem_b = &(_elements[id_b]);
15       }
16       *a = elem_a;
17       *b = elem_b;
18       return;
19   }
```

Fig. 3 Function `netlist::get_random_pair` of *parsec.canneal* that contains two CUs

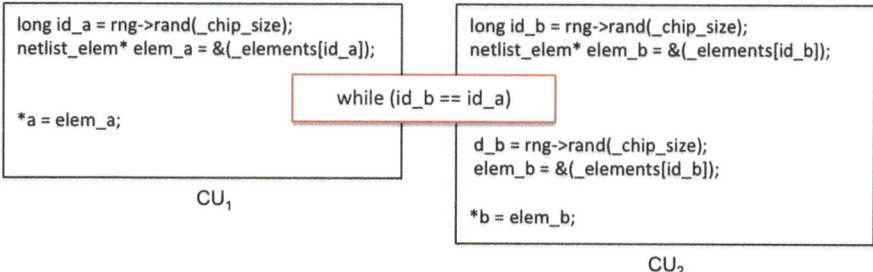

Fig. 4 The two CUs contained in Function `netlist::get_random_pair` of *parsec.canneal*

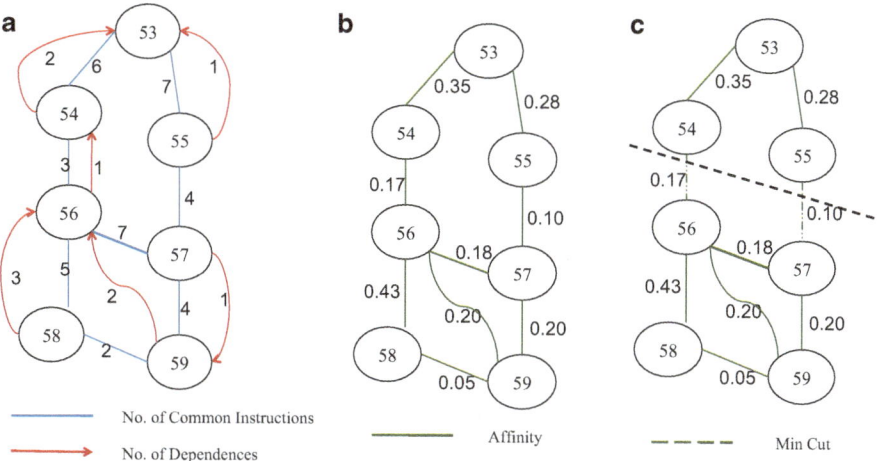

Fig. 5 Example of a CU graph and task formation

store instruction that writes a value to **a* uses all the instructions that correspond to the lines {4, 5, 16} to perform that write operation. Similarly, the group of instructions that correspond to the lines {8, 9, 13, 14, 17} are used for the final *store* instruction that writes **b*. These two sets of instructions can individually be defined as CUs. These CUs form the building blocks of the tasks which can be created for exploiting parallelism in the sequential programs.

Using the common instructions and the RAW dependencies between the CUs, a CU graph is constructed. The nodes of this graph are CU IDs. The CU graph has two types of edges. First type of edge between any two CU nodes signifies RAW dependence between them and is a directed edge. The weight of an RAW edge is the number of RAW dependencies between the two CUs. The second type of edge signifies that there are common instructions between the two CUs. This is an undirected edge and its weight is the number of common instructions between the two CUs. Figure 5a shows a CU graph with red edges as RAW dependencies between the CUs and blue edges as the CUs connected because of the common instructions between them. A CU graph is generally a disconnected graph with several connected components.

3.3   Detecting Parallelism

DiscoPoP finds potential parallelism based on the CU graph. It is well known that among the four kinds of data dependencies, read-after-read (RAR) does not affect parallelization. Write-after-read (WAR) and write-after-write (WAW) are easy to resolve by privatizing the affected variables. Only read-after-write (RAW) seriously prevents parallelization.

3.3.1   DOALL Loops

A loop can be parallelized according to the do-all pattern if there are no loop-carried or inter-iteration dependences. A forward or self-dependence is always loop-carried, as the control flow within a loop iteration moves in a forward direction, which is why dependences within the same iteration must point backward. Note that inner loops in loop nests, which may reverse the control flow direction whenever a new inner iteration starts, are treated separately. The absence of forward or self-dependences is easy to verify based on the graph matrix, whose upper triangle shows all forward and self-edges.

Of course, there may also exist loop-carried dependences in backward edges of the graph matrix. However, to reliably distinguish them from intra-iteration dependences, our dependence profiler would have to record the iteration number along with each memory access, substantially increasing its memory overhead. On the other hand, loop-carried dependences in a backward direction that are not accompanied by dependences in a forward direction in the same loop are very rare. Basically, the absence of forward and self-dependences in a loop is a good indicator of the absence of loop-carried dependences. For this reason, we decided to refrain from the costly classification of backward dependences into loop-carried or not loop-carried.

3.3.2   Tasking

To identify potential parallel tasks, we firstly merge CUs contained in *strongly connected components* (SCCs) or in *chains*. In graph theory, an SCC is a subgraph in which every vertex is reachable from every other vertex. Thus, every CU in an SCC of the CU graph depends on every other CU either directly or indirectly, forming a complex knot of dependences that is likely to defy internal parallelization. Identifying SCCs is important for two reasons:

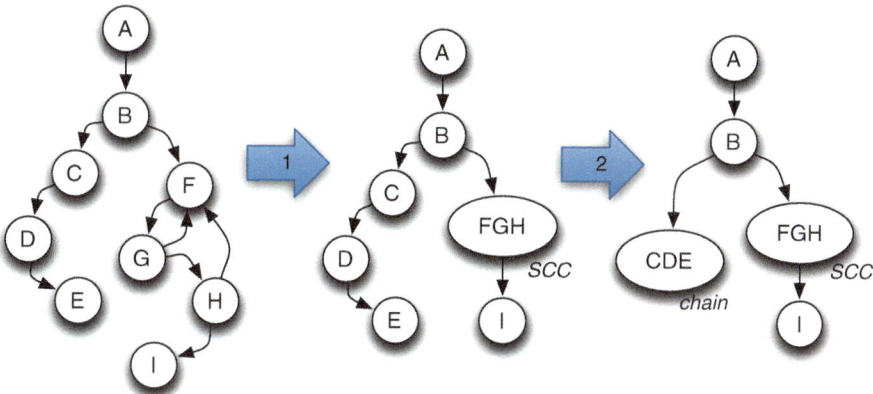

Fig. 6 Forming tasks in a code section of fluidanimate

1. Algorithm design. Complex dependences are usually the result of highly optimized sequential algorithm design oblivious of potential parallelization. In this case, breaking such dependence requires a parallel algorithm, which is beyond the scope of our method.
2. Coding effort. Even if such complex dependences are not created by design, breaking them is usually time-consuming, error-prone, and may cause significant synchronization overhead that may outweigh the benefit of parallelization.

Hence, we hide complex dependences inside SSCs, exposing parallelization opportunities outside, where only a few dependences need to be considered. Figure 6 shows the task-forming process for a code section (starting from serial.c: 341) in function `RebuildGrid()` of fluidanimate, a program from the PARSEC Benchmark Suite [12]. In step 1, CU F, G and H are grouped into SCC_{FGH}. After contracting each SCC to a single vertex, the graph becomes a directed acyclic graph.

Moreover, we group CUs that are connected in a row without a branching or joining point in between into a *chain* of CU. Although they do not form an SCC, we still group them together since each CU contains only a few instructions, and there is no benefit in considering each CU as a separate task. In step 2 of Fig. 6, CU C, D and E are grouped into the $chain_{CDE}$. Finally, we declare each SCC or chain a potential task and derive a parallelization plan from the dependences that exist between them.

After the process of forming chains and SCCs, we can suggest some task parallelism between independent chains and SCCs, that is, without RAW dependencies between them. Note that a chain of CUs may start and end anywhere in the program, without the limitation of predefined constructs, and the code in a chain of CUs does not need to be continuous.

However, some task parallelism can also be utilized with a small amount of refactoring effort, assuming, dependences between potential tasks exist but are weak. We cover these parallelism by applying *minimum cut* on CU graph. In CU

graph, a high value of weight on the edges of any two vertices indicates that those
two CUs either share large amount of computation or they are strongly dependent
on one another. Using these two metrics, we calculate a value called *affinity* for
every pair of CU nodes in the graph. The affinity between any two CU nodes hence
indicates how tightly coupled the two CUs are. A low value of affinity between two
CUs signifies that it's logical to separate the two CUs while forming tasks. The two
types of edges in the graph are replaced by a single undirected edge. The weight of
this edge is the affinity between the two CUs. Figure 5b demonstrates the graph with
the two types of edges between the vertices replaced by single edge with affinity as
the weight.

The next step is to calculate the minimum cut of a connected component using
Stoer-Wagner's algorithm [13]. In graph theory, a *cut* of a graph is a partition of
the vertices of a graph into two disjoint subsets that are joined by at least one edge.
A *minimum cut* is a set of edges that has the smallest number of edges (for an
unweighted graph) or smallest sum of weights possible (for a weighted graph). A
minimum cut creates a disconnected graph with two connected components, each of
which is further analyzed for finding relevant tasks. Figure 5c shows the CU graph
with a minimum cut.

Identifying the minimum cut of a graph divides the graph into two components
that were weakly linked. This indicates that we are separating our code with
minimum number of dependences and common instructions affected. For each
component, the minimum cut is calculated further to divide it into two more
components. The process is repeated recursively over all the components of the
CU graph until the components available are CUs themselves.

3.3.3 Pipeline

To detect pipeline pattern, we use *template-matching* [14] technique. Here, both
template and target program are represented by vectors. Cross-correlation between
two vectors is used to determine how similar they are. We adapt this concept for the
detection of parallel patterns in CU graphs.

Algorithm 1 shows the overall work flow of our approach. We first look for
hotspots in the input program—sections such as loops or functions that have to
shoulder most of the workload. For each hotspot and pattern, we then create a
pattern vector $\mathbf{p}$, whose length is equal to the number n of CUs in the hotspot. The
pattern vector plays the role of the template to be matched to the program. After that,
we create the pattern-specific graph vector $\mathbf{g}$ of the hotspot's CU subgraph, which
represents the part of the program to which the pattern vector is matched. Vectors $\mathbf{p}$
and $\mathbf{g}$ are derived from adjacency matrices reflecting dependences in the pattern and
in the CU graph, respectively. As a next step, we compute the correlation coefficient
of the two vectors using the following formula:

$$CorrCoef = \frac{\mathbf{p} \cdot \mathbf{g}}{\|\mathbf{p}\| \, \|\mathbf{g}\|}$$

Algorithm 1: Parallel pattern detection

$tree = getExecutionTree(serialProgram)$
$Hotspots = findHotspots(tree)$
for *each h in Hotspots* **do**
    $CUGraph = getCUGraph(h)$
    $n = getNumberOfCUs(CUGraph)$
    for *each p in ParallelPatterns* **do**
        $\mathbf{p} = getPatternVector(p, n)$
        $\mathbf{g} = getGraphVector(p, CUGraph)$
        $CorrCoef[h, p] = CorrCoef(\mathbf{p}, \mathbf{g})$
    end
end
$return\ CorrCoef$

The correlation coefficient of the pattern vector and the graph vector of the selected section tells us whether the pattern exists in the selected section or not. The value of the coefficient is always in the range of $[0, 1]$. A 1 indicates that the pattern exists fully, whereas a 0 indicates that it does not exist at all. A value in between shows the pattern can exist but with some limitations which we need to work around. Our tool points out to the dependences, which cause the value of correlation coefficient of a pattern to be less than 1. This helps the programmer to resolve these specific dependences, if he wants to implement that pattern.

Implementing a pipeline only makes sense if its stages are executed many times. For this reason, we restrict our search for pipelines to loops, functions with multiple loops, and recursions. In order to find a pipeline, we first let DiscoPoP deliver the CU graphs of all hotspots. Because DiscoPoP counts the number of read and write instructions executed in each loop or function, we currently use this readily available metric as an approximation of the workload when searching for hotspots. A more comprehensive criterion, including execution times and workload, will be implemented in the future. We then compute an adjacency matrix for each hotspot graph, which we call the *graph matrix*. For each graph matrix, we create a corresponding pipeline pattern matrix of the same size, which we call the *pipeline matrix*. Pipeline matrices encode a very specific arrangement of dependences expected between CUs. For example, there must be a dependence chain running through all CUs in the graph because a pipeline consists of a chain of dependent stages. This specific property helps us to derive the pipeline pattern vector from the matrix.

4 Evaluation and Results

We conducted a range of experiments to evaluate the ability of DiscoPoP to detect DOALL loops, potential parallel tasks, and parallel patterns. The performance of DiscoPoP is also analyzed. Test cases are the NAS Parallel Benchmarks 3.3.1 [6]

(NAS), a suite of programs derived from real-world computational fluid-dynamics applications, the Starbench parallel benchmark suite [15] (Starbench), which covers programs from diverse domains, including image processing, information security, machine learning and so on, benchmarks from PARSEC Benchmark Suite 3.0 [12], and a few real-world applications. Whenever possible, we tried different inputs to compensate for the input sensitivity of dynamic dependence profiling.

4.1 DOALL Loops

To evaluate the ability of DiscoPoP to detect DOALL loops, we searched for parallelizable loops in sequential NPB programs and compared the results with the parallel versions provided by NPB. Table 1 shows the results of the experiment. The data listed in the column set "Executed" are obtained dynamically. Column "# loops" gives the total number of loops which were actually executed. The number of loops that we identified as parallelizable are listed under "# parallelizable". At this stage, prior to the ranking, DiscoPoP considers only data dependencies, which is why many loops carrying no dependency but bearing only a negligible amount of work are reported. The second set of columns shows the number of annotated loops in OpenMP versions of the programs (# OMP). Under "# identified" we list how many annotated loops were identified as parallelizable by DiscoPoP.

As shown in Table 1, DiscoPoP identified 92.5 % (136/147) of the annotated loops. A comparison with other tools is challenging because none of them is available for download. A comparison based exclusively on the literature has to account for differences in evaluation benchmarks and methods. For Parwiz [2], the authors reported an average of 86.5 % after applying their tool to SPEC OMP-2001. Kremlin [4], which was also evaluated with NPB, selects only loops whose expected speedup is high. While Kremlin reported 55.0 % of the loops annotated in NPB, the top 30 % of DiscoPoP's ranked result list cover 65.3 % (96/147).

Table 1 Detection of parallelizable loops in NAS Parallel Benchmark programs

Program	Executed		OpenMP-annotated loops			
	# loops	# parallelizable	# OMP	# identified	# in top 30 %	# in top 10 %
BT	184	176	30	30	22	9
SP	252	231	34	34	26	9
LU	173	164	33	33	23	7
IS	25	20	11	8	2	2
EP	10	8	1	1	1	1
CG	32	21	16	9	5	5
MG	74	66	14	14	11	7
FT	37	34	8	7	6	5
Overall	787	720	147	136	96	45

4.2 Tasking

We have applied two separate strategies to evaluate the ability of DiscoPoP to detect parallel tasks. Firstly, we compare the parallel implementations of the applications from the Starbench benchmark suite with the tasks identified by our analysis. In this case, our goal is to verify if the approach identifies valuable and logical homogeneous tasks. Secondly, we use some of the programs from the PARSEC benchmark and parallelize these applications based on the heterogeneous tasks which are identified as potential candidates for parallelism.

4.2.1 Comparison with Existing Parallel Implementations

Our first evaluation strategy involves providing a comparison of the identified tasks with the existing parallel versions of the applications for Starbench parallel benchmark suite. The Table 2 shows the overview of the evaluation performed. Column "Task suggestion" shows the location where parallel tasks are identified using our approach and column "Matched in parallel implementations" shows whether the identified tasks exist in the official parallel implementations from Starbench. The tasks were identified by prioritizing the main algorithm functions and the functions that consumed the majority of the total execution time of the program as shown in column "Execution time (%)".

As an example, we show the tasks identified in *k-means*, a clustering algorithm widely used in the domains of data-mining and artificial intelligence. The application consists of two iteratively repeated phases. One is a clustering phase and the other is a reduction phase that computes new clusters. In the sequential version, the function kmeans() calls the function cluster() which performs the clustering phase. The remaining body of the function kmeans() performs the reduction phase. The function cluster() takes 99.6 % of the total execution time of the program. This makes it a good candidate for analysis of the tasks from the list of the tasks identified for the program.

Table 2 Parallel tasks identified in Starbench compared to existing parallel implementations

Program	Task suggestion	Execution time (%)	Matched in parallel implementations	# CUs used
c-ray	render_scanlines()	100.0	Yes	4
k-means	cluster()	99.6	Yes	3
md5	MD5_Update()	93.5	Yes	7
rotate	RotateEngine::run()	90.3	Yes	6
rgbyuv	processImage()	100.0	Yes	7
ray-rot	render_scanlines()	97.2	Yes	10
rot-cc	RotateEngine::run()	54.7	Yes	13

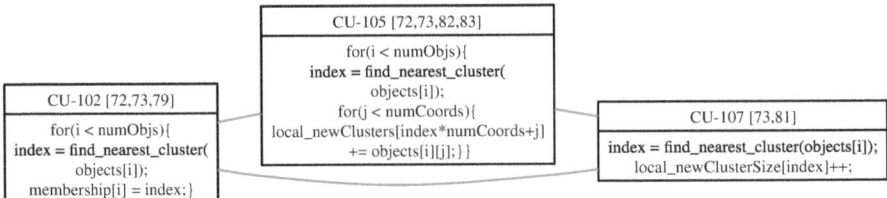

Fig. 7 Connected component of the CU graph of *k-means* corresponding to function `cluster()`

Table 3 Speedups obtained by parallelizing identified tasks in PARSEC benchmarks

Program	Function	Code refactoring	# Threads	Local speedup
Fluidanimate	`RebuildGrid()`	Yes	2	1.60
Fluidanimate	`ProcessCollisions()`	No	4	1.81
Canneal	`routing_cost_given_loc()`	Yes	2	1.32
Blackscholes	`CNDF()`	NA	NA	NA

The analysis identifies both of the aforementioned phases individually as tasks. The `cluster()` function is identified as a task by grouping 3 CUs from the CU graph. Figure 7 shows the connected component of the graph for the function `cluster()`. As for the reduction phase, only the part of the function `kmeans()` that performs this phase is identified as task by the analysis. In the pthreads version of the program, every thread executes the function `work()`, which contains the same code as the sequential version of `cluster()`. The reduction phase is run by the main thread thereafter.

4.2.2 Parallelization Based on the Suggested Heterogeneous Tasks

In this section we investigate some applications of PARSEC benchmark suite. We parallelized these applications based on the tasks formed by using CUs or by directly considering CUs as tasks. We assigned these tasks to separate threads and calculated the speedup obtained. We parallelized these cases mainly using OpenMP `section` and `task` directives. Table 3 shows the results of the applications parallelized. The local speedups represent an average of five independent executions of the programs. Column "# Threads" shows the number of threads used to parallelize the suggestions. Column "Code refactoring" indicates if the refactoring the code like adding necessary synchronization or replicating some part of the code across multiple threads was necessary to parallelize the program based on the suggestion. For our future work, we would like to predict the various kinds of synchronizations or code refactoring necessary to parallelize a suggestion based on the available dependences and CUs identified.

4.3 Pipeline

We applied DiscoPoP on five benchmarks from PARSEC and libVorbis to detect pipeline pattern. Among the test cases, three benchmarks (*bodytrack*, *dedup*, and *ferret*) have pipeline patterns according to their existing parallel implementations, while two of the test cases (*blackscholes* and *fluidanimate*) do not. libVorbis has a natural pipeline work flow, but we cannot confirm it before applying DiscoPoP since there is no existing parallel implementation for it.

Table 4 shows the results of detecting pipeline patterns on the test cases. For all the three cases that contain pipeline patterns in parallel implementations, the utilized pipeline are successfully identified. Thus we did not parallelize them again. No pipeline pattern is detected in *blackscholes*, which is also as expected. However, pipeline pattern is detected in *fluidanimate*, which does not have a pipeline in its parallel implementations. After examining the code, we believe the parallelism does exist, and we parallelized the code section following the suggestion. Our parallelization yields a speedup of 1.52 using three threads. Note that the correlation coefficient for this pipeline pattern is 0.94, which implies that code refactoring may be needed. Actually, parallelizing this place requires quite a lot of effort.

LibVorbis is a reference implementation of the Ogg Vorbis codec. It provides both a standard encoder and decoder for the Ogg Vorbis audio format. In this study, we analyzed the encoder part. The suggested pipeline resides in the body of the loop that starts at file encoder_example.c, line 212, which is inside the main function of the encoder. The pipeline contains only two stages: `vorbis_analysis()`, which applies some transformation to audio blocks according to the selected encoding mode (this process is called analysis), and the remaining part that actually encodes the audio block. After investigating the loop of the encoding part further, we found it to have two sub-stages: encoding and output.

We constructed a four-stage pipeline with one stage each for analysis, encoding, serialization, and output, respectively. We added a serialization stage, in which we reorder the audio blocks because we do not force audio blocks to be processed in order in the analysis and the encoding phase. We ran the test using a set of uncompressed wave files with different sizes, ranging from 4 to 47 MB. As a result, the parallel version achieved an average speedup of 3.62 with four threads.

Table 4 Pipeline patterns identified in PARSEC benchmarks and libVorbis

Program	# of pipeline in parallel version	Corr. coef.	# Detected	Speedup
Bodytrack	1	0.96	1	N.A.
Dedup	1	1.00	1	N.A.
Ferret	1	1.00	1	N.A.
Blackscholes	0	0.00	0	N.A.
Fluidanimate	0	0.94	1	1.52 (3T)
LibVorbis	N.A.	1.00	1	3.62 (4T)

4.4   Overhead

We conducted our performance experiments on a server with 2×8-core Intel Xeon E5-2650 2 GHz processors with 32 GB memory, running Ubuntu 12.04 (64-bit server edition). All the test programs were compiled with option -g -O2 using Clang 3.3. For NAS, we used the input set W; for Starbench, we used the reference input set.

4.4.1   Time Overhead

First, we examine the time overhead of our profiler. The number of threads for profiling is set to 8 and 16. The slowdown figures are average values of three executions compared with the execution time of uninstrumented runs. The negligible time spent in the instrumentation is not included in the overhead. For NAS and Starbench, instrumentation was always done in 2 s.

The slowdown of our profiler when profiling sequential programs is shown in Fig. 8. The average slowdowns for the two benchmark suites ("NAS-average" and "Starbench-average") are also included. As the figure shows, our serial profiler has a 190× slowdown on average for NAS benchmarks and a 191× slowdown on average for Starbench programs. The overhead is not surprising since we perform an exhaustive profiling for the whole program.

When using 8 threads, our parallel profiler gives a 97× slowdown (best case 19×, worst case 142×) on average for NAS benchmarks and a 101× slowdown (best case 36×, worst case 253×) on average for Starbench programs. After increasing the number of threads to 16, the average slowdown is only 78× (best case 14×, worst case 114×) for NAS benchmarks, and 93× (best case 34×, worst case 263×) for Starbench programs. Compared to the serial profiler, our parallel profiler achieves a 2.4× and a 2.1× speedup using 16 threads on NAS and Starbench benchmark suites, respectively.

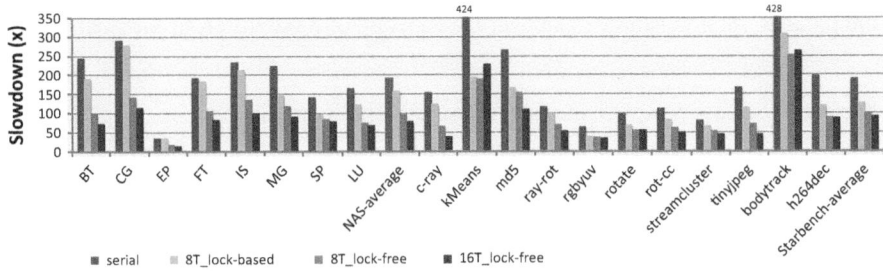

Fig. 8 Slowdowns of data dependence profiler for sequential NAS and Starbench benchmarks

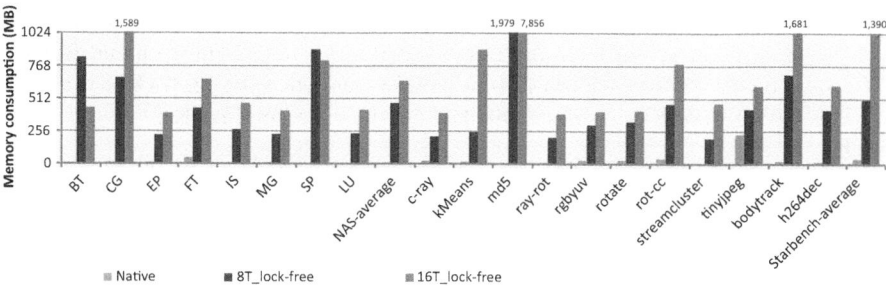

Fig. 9 Memory consumption of the profiler for sequential NAS and Starbench benchmarks

4.4.2 Memory Consumption

We measure memory consumption using the "maximum resident set size" value provided by /usr/bin/time with the verbose (-v) option. Figure 9 shows the results. When using 8 threads, our profiler consumes 473 MB of memory on average for NAS benchmarks and 505 MB of memory on average for Starbench programs. The average memory consumption is increased to 649 and 1,390 MB for NAS and Starbench programs, respectively. The worst case happens when using 16 threads to profile *md5*, which consumes about 7.6 GB memory. Although this may exceed the memory capacity configured in a three-year-old PC, it is till adequate for up-to-date machines, not to mention servers that are usually configured with 16 GB memory or more.

5   Conclusion and Outlook

We introduced a novel dynamic tool for the discovery of potential parallelism in sequential programs. By building the concept of computational units (CUs) and embedding it in a framework of combined static and dynamic analysis, we can reconcile the identification of parallel tasks in the form of CU chains with efficiency both in terms of time and memory. CU chains are not confined to predefined language constructs but can spread across the whole program. Our approach found 92.5 % of the parallel loops in NAS Parallel Benchmark (NPB) programs and successfully identified tasks spanning several language constructs as well as pipeline patterns. It also helped parallelizing a loop in spite of initial dependences. Implementing the generated plan achieved reasonable speedups in most of our test cases: up to 2.67 for independent tasks and up to 3.62 for pipelines using a maximum of four threads.

In the future, we want to support further types of task parallelism including, for example, TBB flow graph. Furthermore, we want to develop heuristics to validate the suggestions before submitting them to the programmer, providing more accurate and reliable results.

References

1. Johnson, R.E.: Software development is program transformation. In: Proceedings of the FSE/SDP Workshop on Future of Software Engineering Research, FoSER'10, Santa Fe, pp. 177–180. ACM (2010)
2. Ketterlin, A., Clauss, P.: Profiling data-dependence to assist parallelization: framework, scope, and optimization. In: Proceedings of the 45th Annual IEEE/ACM International Symposium on Microarchitecture, MICRO 45, Vancouver, pp. 437–448. IEEE Computer Society (2012)
3. Zhang, X., Navabi, A., Jagannathan, S.: Alchemist: a transparent dependence distance profiling infrastructure. In: Proceedings of the 7th Annual IEEE/ACM International Symposium on Code Generation and Optimization, CGO'09, Seattle, pp. 47–58. IEEE Computer Society (2009)
4. Garcia, S., Jeon, D., Louie, C.M., Taylor, M.B.: Kremlin: rethinking and rebooting gprof for the multicore age. In: Proceedings of the 32nd ACM SIGPLAN Conference on Programming Language Design and Implementation, PLDI'11, San Jose, pp. 458–469. ACM (2011)
5. Li, Z., Jannesari, A., Wolf, F.: Discovery of potential parallelism in sequential programs. In: Proceedings of the 42nd International Conference on Parallel Processing, PSTI'13, Lyon, pp. 1004–1013. IEEE Computer Society (2013)
6. Bailey, D.H., Barszcz, E., Barton, J.T., Browning, D.S., Carter, R.L., Fatoohi, R.A., Frederickson, P.O., Lasinski, T.A., Simon, H.D., Venkatakrishnan, V., Weeratunga, S.K.: The NAS parallel benchmarks. Int. J. Supercomput. Appl. **5**(3), 63–73 (1991)
7. Kim, M., Kim, H., Luk, C.K.: SD3: a scalable approach to dynamic data-dependence profiling. In: Proceedings of the 43rd Annual IEEE/ACM International Symposium on Microarchitecture, MICRO 43, Atlanta, pp. 535–546. IEEE Computer Society (2010). http://www.microarch.org/micro43/
8. Moseley, T., Shye, A., Reddi, V.J., Grunwald, D., Peri, R.: Shadow profiling: hiding instrumentation costs with parallelism. In: Proceedings of the 5th International Symposium on Code Generation and Optimization, CGO'07, San Jose, pp. 198–208. IEEE Computer Society, Washington, DC (2007)
9. Kim, M., Kim, H., Luk, C.K.: Prospector: discovering parallelism via dynamic data-dependence profiling. In: Proceedings of the 2nd USENIX Workshop on Hot Topics in Parallelism, HOTPAR'10, Berkeley (2010)
10. Amini, M., Goubier, O., Guelton, S., Mcmahon, J.O., Xavier Pasquier, F., Pan, G., Villalon, P.: Par4All: from convex array regions to heterogeneous computing. In: Proceedings of the 2nd International Workshop on Polyhedral Compilation Techniques, IMPACT 2012, Paris (2012)
11. Grosser, T., Groesslinger, A., Lengauer, C.: Polly – performing polyhedral optimizations on a low-level intermediate representation. Parallel Process. Lett. **22**(04), 1250010 (2012)
12. Bienia, C.: Benchmarking modern multiprocessors. Ph.D. thesis, Princeton University (2011)
13. Von Luxburg, U.: A tutorial on spectral clustering. Stat. Comput. **17**(4), 395–416 (2007)
14. Dong, J., Sun, Y., Zhao, Y.: Design pattern detection by template matching. In: Proceedings of the 2008 ACM Symposium on Applied Computing, SAC'08, Fortaleza, pp. 765–769. ACM (2008)
15. Andersch, M., Juurlink, B., Chi, C.C.: A benchmark suite for evaluating parallel programming models. In: Proceedings 24th Workshop on Parallel Systems and Algorithms, PARS'11, Rüschlikon, pp. 7–17 (2011)

Tareador: The Unbearable Lightness of Exploring Parallelism

Vladimir Subotic, Arturo Campos, Alejandro Velasco, Eduard Ayguade, Jesus Labarta, and Mateo Valero

Abstract The appearance of multi/many-core processors created a gap between the parallel hardware and sequential software. Furthermore, this gap keeps increasing, since the community cannot find an appealing solution for parallelizing applications. We propose Tareador as a mean for fighting this problem.

Tareador is a tool that helps a programmer explore various parallelization strategies and find the one that exposes the highest potential parallelism. Tareador dynamically instruments a sequential application, automatically detects data-dependencies between sections of execution, and evaluates the potential parallelism of different parallelization strategies. Furthermore, Tareador includes the automatic search mechanism that explores parallelization strategies and leads to the optimal one. Finally, we blueprint how Tareador could be used together with the parallel programming model and the parallelization workflow in order to facilitate parallelization of applications.

1 Introduction

Parallel programming became an urge, an urge that the programmers community fails to efficiently respond to. One of the biggest problems in the current computing industry is the steady-growing gap between the increasing parallelism offered by state-of-the-art architectures and the limited parallelism exposed in state-of-the-art applications. Consequently, software parallelism has become concern of every single programmer. However, parallelizing applications is far from trivial.

The community keeps inventing novel programming models as enablers for transition to parallel software. These novel programming models come in various flavours, offering different programming paradigms, levels of abstraction, etc. However, most of the novel programming models fail to get widely adopted. It takes a giant leap of faith for a programmer to take the already working application and to port it to a novel programming model. This is especially problematic because the

V. Subotic (✉) • A. Campos • A. Velasco • E. Ayguade • J. Labarta • M. Valero
Barcelona Supercomputing Center, Barcelona, Spain
e-mail: vladimir.subotic@bsc.es; arturo.sanemeterio@bsc.es; alejandro.velasco@bsc.es;
eduard.ayguade@bsc.es; jesus.labarta@bsc.es; mateo@ac.upc.edu

C. Niethammer et al. (eds.), *Tools for High Performance Computing 2014*,
DOI 10.1007/978-3-319-16012-2_4

55

Fig. 1 There is a need to make the complete parallelization solution that will contain not only the development tool (*DT*), but also the proposal of the programming model (*PM*) and the description of workflow (*WF*) that guides the process of parallelizing applications

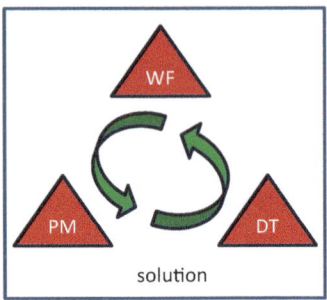

programmer cannot anticipate how would the application perform within the new programming model and thus whether the porting is worth the effort. Moreover, the programmer usually lacks development tools and a clear idea how the parallelization process should be conducted.

It is our belief that the programmers should be offered not only with the programming model, but with the whole parallelization solution – a solution that includes parallel programming model, parallelization development tool, and the parallelization workflow that describes how to use the development tool to port the sequential application to the selected parallel programming model. We believe that the three parts of the solution need to be tailored to work together in the bigger system (Fig. 1). The development tool should instrument the sequential application and provide to the user the information that is relevant in the context of the target parallel programming model. Finally, the parallelization workflow should glue the tool and the programming model and define the process of parallelizing applications.

This paper attempts to paint the big picture of parallelization solution, putting the special emphasize on the part of the parallelization development tool. More specifically, this paper contributes in the following two directions:

- We present Tareador – a tool for assisted parallelization of sequential applications. Tareador allows the programmer to understand the inner workings of the application, identify the dependencies between different parts of the execution and evaluate the parallelism inherent in the code. We describe Tareador at its current development phase, as a tool to explore potential parallelization strategies. Furthermore, we also discuss the planed future development of Tareador in an effort to make it a complete tool for assisted parallelization of applications.
- We describe the parallelization solution that includes Tareador. We propose a parallel programming model that best suits the Tareador obtained information, and describe the workflow that uses Tareador to parallelize application by porting it to the target programming model.

The rest of the paper is organized as follows. Section 2 illustrates the problem of finding the optimal parallelization strategy. Section 3 describes the implementation and usage of Tareador environment. Furthermore, in Sect. 4 we describe the automatic algorithm that automatically drives Tareador in exploration of good parallelization strategies. We describe the heuristics and metrics that guide the automatic search and show the results of automatic search in the case of couple of well-known applications. Furthermore, we include a broad discussion of how we see Tareador being a part of the whole environment for easy parallelization of applications (Sect. 5). We declare our selection of the parallel programming model and devise a custom parallelization workflow that would facilitate parallelization of applications. Finally, we conclude the paper with the related work on the topic (Sect. 6) and the conclusions of our study (Sect. 7).

2 Motivating Example

Parallelization of a sequential application consists of decomposing the code into tasks (e.g. units of parallelism) and implementing synchronization rules between the created tasks. However, even if the sequential application is simple, finding the optimal task decomposition can be a difficult job. The application may exhibit parallelism that is very distant and irregular, parallelism among sections of code that are mutually far from each other. This type of parallelism is very hard for the programmer to identify and expose without any development support. Thus, to find the optimal parallelization strategy, the programmer must know the source code in depth in order to identify all the data dependencies among tasks. Furthermore, the programmer must anticipate how will all the tasks execute in parallel, and what is the possible parallelism that these tasks can achieve.

Figure 2 shows a simple sequential application composed of four computational parts, the data dependencies among those parts, and some of the possible taskification strategies. Although the application is very simple, it allows various decompositions that expose different amount of parallelism. $T0$ puts all the code in one task and, in fact, presents a sequential code. $T1$ and $T2$ both break the application into two tasks but fail to expose any parallelism. On the other hand, $T3$ and $T4$ both break the application into 3 tasks, but while $T3$ achieves no parallelism, $T4$ exposes concurrency between C and D. Finally, $T5$ breaks the application into 4 tasks but achieves the same amount of parallelism as $T4$. Considering that increasing the number of tasks increases the runtime overhead, one can conclude

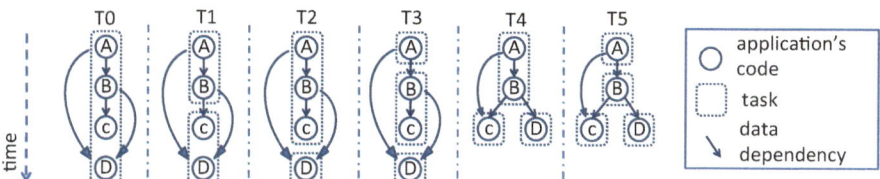

Fig. 2 Execution of different possible taskifications for a code composed of four parts

that the optimal taskification is $T4$, because it gives the highest speedup with the lowest cost of the increased number of tasks.

Nevertheless, compared to the presented trivial execution, a real-world application would be more complex in various aspects. A real application may have hundreds of thousands of task instances, causing complex and well populated dependency graphs. The large dependency graph would allow unpredictable scheduling decisions that would potentially exploit distant parallelism. Also, with the task instances of different duration, evaluating the potential parallelism would be even harder. Due to all this complexity, it is unfeasible for a programmer alone to do the described analysis and estimate the potential parallelism of a certain task decomposition. Therefore, we believe that it would be very useful to have an environment that quickly anticipates the potential parallelism of a particular taskification. We describe such a framework in the following section.

3   Tareador Environment

Tareador allows the programmer to start from a sequential application, propose some decomposition of the sequential code into tasks and get fast estimation of the potential parallelism. The input to Tareador is a sequential code. Tareador compiler marks all logical sections of code as potential tasks. In addition, the user can manually annotate other potential task. The annotated code is executed sequentially – all annotated tasks are executed in the order of their instantiation. Tareador dynamically instruments the sequential execution and collects the log of memory usage of each potential task. Once the logs are generated, Tareador allows the programmer to select one task decomposition of the sequential code. For the selected decomposition, Tareador calculates inter-task dependencies and evaluates the potential parallelism of the decomposition providing to the user the results in the form of:

- Simulation of the potential parallel execution;
- Dependency graph of all task instances;
- Visualization of the memory usage of each task.

Tareador environment integrates various internally and externally developed tools. The framework (Fig. 3) takes the **input code** and compiles it with LLVM-based [1] **Tareador compiler**. The execution of the obtained binary generates Tareador **execution logs**. Further post-mortem processing of the execution logs is encapsulated into **Tareador GUI**. Tareador GUI allows the user to select one decomposition. Based on the selected task decomposition, **Tareador backend** consumes the execution logs to calculate final results. More specifically, Tareador backend generates execution trace that **Dimemas** [2] simulates to obtain **Paraver** [3] time-plots of the potential parallel execution. Also, Tareador backend produces the task dependency graph that can be visualized with **Graphviz** [4], as well as dataview information that can be visualized internally by Tareador GUI.

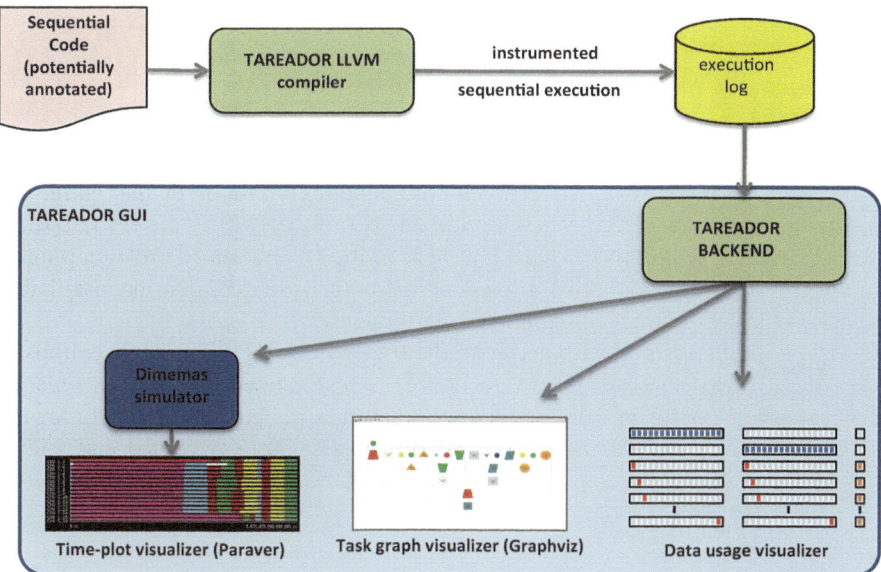

Fig. 3 Tareador framework

3.1   Implementation Details

Tareador uses LLVM framework to dynamically instrument the sequential application and collect the log of all potential tasks and their memory usage. Tareador compiler injects to the original sequential execution instrumentation callbacks that collect the data needed for Tareador analysis. First, Tareador compiler must mark all the potential tasks in the execution. The compiler marks as a potential task every logical code section that can take a significant amount of time – each function, loop or loop iteration. Also, the compiler allows the user to manually annotate any potential task by wrapping an arbitrary code sections using Tareador API (example in Sect. 3.2). Furthermore, Tareador intercepts and processes each memory access of the sequential execution. Finally, based on the dynamically collected information, Tareador flushes the execution log that contains all intercepted potential tasks and their memory usage. The resulting log is indexed to allow fast post-mortem browsing.

Tareador GUI allows the user to easily browse different task decompositions of the instrumented execution. Given the configuration of one task decomposition, Tareador backend consumes the execution logs to evaluate the parallelism of the decomposition. The backend finds all the specified tasks in the execution log, calculates data-dependencies between them and prepares outputs for different visualization tools. Finally, GUI allows the user to see all the obtained results and select how to refine the decomposition to achieve higher parallelism.

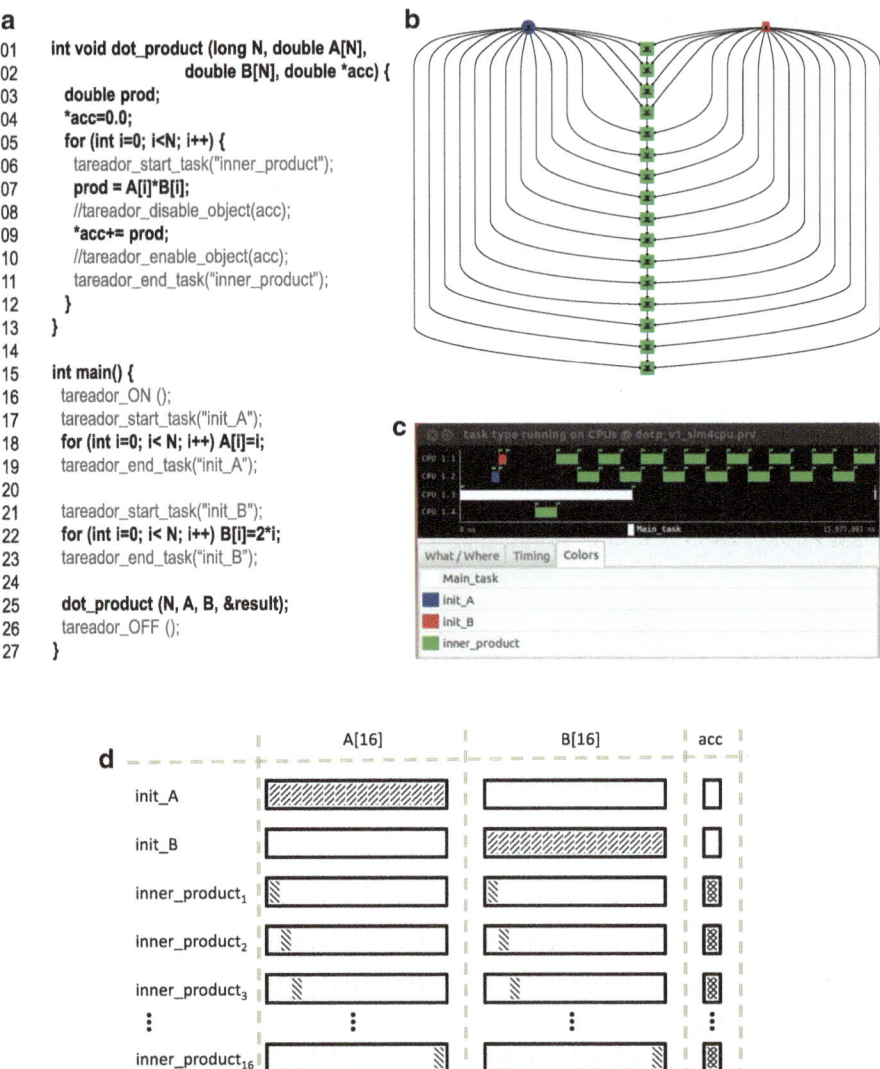

a

```
01    int void dot_product (long N, double A[N],
02                          double B[N], double *acc) {
03      double prod;
04      *acc=0.0;
05      for (int i=0; i<N; i++) {
06        tareador_start_task("inner_product");
07        prod = A[i]*B[i];
08        //tareador_disable_object(acc);
09        *acc+= prod;
10        //tareador_enable_object(acc);
11        tareador_end_task("inner_product");
12      }
13    }
14
15    int main() {
16      tareador_ON ();
17      tareador_start_task("init_A");
18      for (int i=0; i< N; i++) A[i]=i;
19      tareador_end_task("init_A");
20
21      tareador_start_task("init_B");
22      for (int i=0; i< N; i++) B[i]=2*i;
23      tareador_end_task("init_B");
24
25      dot_product (N, A, B, &result);
26      tareador_OFF ();
27    }
```

Fig. 4 Applying Tareador on dot product kernel. (**a**) Source code. (**b**) Task dependency graph. (**c**) Potential parallel execution (4 cores). (**d**) Visualization of memory accesses

3.2 Illustration of Tareador Usage

This section illustrates the usage of Tareador by applying it on a simple code of dot product computation (Fig. 4). Figure 4a shows the original sequential code (code in black). The code initializes two buffers operands in two loops, and then computes the result in function *dot_product*. In order to prepare the execution for Tareador instrumentation, the user adds the gray code lines. To mark which code section will be instrumented by Tareador, the user inserts functions *tareador_ON* and *tareador_OFF* (code lines 16 and 26). Furthermore, to propose one task decomposition, the user inserts calls *tareador_start_task* and *tareador_end_task*. Additional strings passed to these functions mark the name of the task that is encapsulated by the matching calls. In the presented examples, the selected task decomposition splits the sequential execution into 2 initializing tasks (*init_A* and *init_B*) and 16 computational tasks (*inner_product*), one for each iteration of the loop in *dot_product* function. It is important to note that Tareador can work without these user annotations that mark the decomposition. Tareador compiler can automatically mark all the potential tasks and then the different task-decomposition could be browsed through Tareador GUI, without the need to modify the target sequential code.

For the selected task decomposition, Tareador automatically evaluates the potential parallelism. First output that evaluates parallelism is a tasks dependency graph (Fig. 4b). The tasks dependency graph is a directed cyclic graph where each node represents a task instance, while each edge represents a data-dependency between two task instances. In the presented example, the blue and red nodes represent tasks *init_A* and *init_B*, while the green nodes represent instances of *inner_product*. The graph shows that each instance of *inner_product* depends on *init_A*, *init_B* and the previous instance of *inner_product* (if any). The second Tareador output is the time-plot of the potential parallel execution of the selected decomposition (Fig. 4c). The figure shows, for each of the 4 cores in the parallel machine (y-axis), which task executes in any moment of time (x-axis). The colors representing task types match the colors from the dependency graph. The presented plot confirms that green task instances (*inner_product*) are serialized.

Tareador's dataview visualization can further pinpoint the memory objects that impede parallelism. Figure 4d shows for all task instances the memory access patterns within the objects of interest. As expected, the initialization tasks (*init_A* and *init_B*) access only their target arrays. On the other hand, each instance of *inner_product* reads one element from both arrays *A* and *B* and increments *acc* (inout access stands for both input and output). Therefore, dependencies between instances of *inner_product* are caused by the memory object *acc*.

Going back to the source code for from Fig. 4a, we can recognize the dependency on the object *acc* as an apparent case that can be avoided using reduction. Thus, Tareador allows the user to evaluate the potential parallelism if the dependency on *acc* is to be avoided using reduction. The user can uncomment the code lines 8 and 10 and declare that, within the encapsulated code snippet, the memory accesses to *acc* should be omitted. In other words, the user instructs Tareador to ignore the dependency on the object *acc*. Consequently, the resulting decomposition allows concurrency between instances of *inner_product*, as shown in Fig. 5.

For the illustration purpose, in this section we decided to describe the usage of the **lite** mode of Tareador. The lite mode requires the user to manually mark the task decomposition of the code. Every time the user specifies the decomposition,

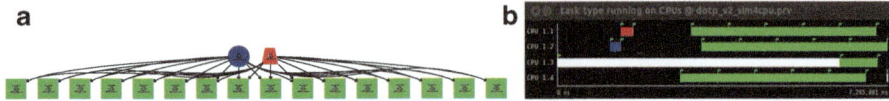

Fig. 5 Applying Tareador on dot product kernel (disabled accesses to *acc*). (**a**) Task dependency graph. (**b**) Potential parallel execution (4 cores)

Tareador outputs the described results. Conversely, the **original** Tareador mode requires no annotations of the target sequential code. In the original mode, Tareador compiler automatically marks all the potential tasks in the sequential code, and lets the user browse all the potential decompositions through Tareador GUI. For the purpose of parallelizing applications, original mode is much more efficient than then lite mode. However, for demonstrative/teaching purposes, the lite mode is preferred. This is because the lite mode makes the students be more involved with the actual target code. Every time the student wants to change the decomposition, she is forced to interact with the target code and understand better the sources of parallelism. The lite mode of Tareador has been successfully introduced into the teaching curriculum of parallel programming courses at the Technical University of Catalonia.

4   Automatic Exploration of Parallelism with Tareador

In our prior work [5] we demonstrated how a programmer can use Tareador to iteratively explore the task decomposition space and find the decomposition that exposes sufficient parallelism to efficiently deploy multi-core processors. However, the presented process relied strongly on programmer's experience to guide the search. To further facilitate the process of finding optimal parallelization strategy, our next goal is to formalize the programmers' experience into an autonomous algorithm for automatic search of potential parallelization strategies. The rest of this section describes the autonomous algorithm and metrics and heuristics that define it.

The automatic exploration of parallelization strategies is based on: evaluating parallelism of various decompositions; collecting key parameters that identify the parallelization bottlenecks; and refining decompositions in order to increase parallelism. The search algorithm is illustrated in Fig. 6. The inputs of this algorithm are the original unmodified sequential code and the number of cores in the target platform. The search algorithm passes through the following steps:

1. Start from the most coarse-grain task decomposition, i.e. the one that considers the whole main function as a single task.
2. Perform an estimation of the potential parallelism of the current task decomposition (the speedup with respect to the sequential execution).

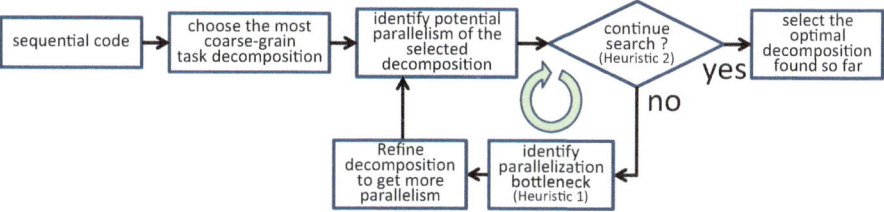

Fig. 6 Algorithm for exploring parallelization strategies

3. If the exit condition is met (*Heuristic 2*), finish the search.
4. Else, identify the parallelization bottleneck (*Heuristic 1*), i.e. the task that should be decomposed into finer-grain tasks.
5. Refine the current task decomposition in order to avoid the identified bottleneck. Go to step 2.

4.1 Algorithm Heuristics

In the following sections, we further describe the design choices made in designing the mentioned heuristics. Nevertheless, first we must define more precise terminology. Primarily, we must make a clear distinction between a **task type** (function that is encapsulated into task) and a **task instance** (dynamic instance of that function). For instance, if function *compute* is encapsulated into a task, we will say that *compute* is a **task type**, or just a **task**. Conversely, each instantiation of *compute* we will call a **task instance**, or just an **instance**. A task instance is atomic and sequential, but various instances (of same or different task type) can execute concurrently among themselves.

Also, we will often use a term **breaking a task** to refer to the process of transforming one task into more fine-grain tasks. For example, Fig. 7 illustrates decomposition refining in a case of a simple code. The process starts with the most coarse-grain decomposition (*D1*) in which function *A* is the only task. By breaking task *A*, we obtain decomposition *D2* in which *A* is not a task and instead its direct children (*B* and *C*) become tasks. If in the next step we break task *B*, assuming that *B* contains no children tasks, *B* will be serialized (i.e. *B* is not a task anymore and its computation becomes a part of the sequential execution). Similarly, the next refinement serializes task *C* and leads to the starting sequential code. At this point, no further refinement is possible, so the iterative process naturally stops.

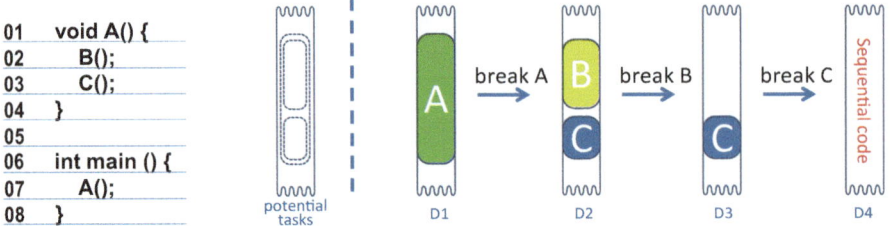

```
01    void A() {
02        B();
03        C();
04    }
05
06    int main () {
07        A();
08    }
```

Fig. 7 Iterative refinement of decompositions

4.1.1   Heuristic 1: Which Task to Break

In the manual search for an efficient decomposition, the programmer decides which task is the parallelization bottleneck. The practice shows that the bottleneck task is often one of the following:

1. **The task whose instances have long duration**, because a long instance may cause significant load imbalance.
2. **The task whose instances have many dependencies**, because an instance with many dependencies may be a strong synchronization point.
3. **The task whose instances have low concurrency**, because an instance with low concurrency may prevent other instances to execute in parallel.

Our goal is to formalize this programmer experience into a simple set of metrics that can lead an autonomous algorithm for exploring potential task decompositions. The goal is to define a **cost function** for task type i as:

$$\overline{t_i} = \overline{l_i(p_l)} + \overline{d_i(p_d)} + \overline{c_i(p_c)} \tag{1}$$

where l_i, d_i and c_i are functions that calculate the partial costs related to tasks' length, dependencies count and concurrency level. On the other hand, parameters p_l, p_d and p_c are empirically identified parameters that tune the weight of each partial cost within the overall cost. The following paragraphs further describe the operands from Eq. 1.

Metric 1: Task Length Cost

A task type that has long instances is a potential parallelization bottleneck. Thus, based on the length of instances, we define a metric called length cost of a task type. Length cost of some task type is proportional to the length of the longest instance of that task. Therefore, if task i has instances whose lengths are in the array T_i, the length cost of task i is:

$$l_i = \max(t), t \in T_i \tag{2}$$

Furthermore, we define a normalized length cost of task i as:

$$\overline{l_i(p)} = \frac{(l_i)^p}{\sum\limits_{j=1}^{N} (l_j)^p}, \quad 0 \leq p < \infty \tag{3}$$

where the control parameter p is used to tune the distribution of normalized costs (explained later in this section).

Metric 2: Task Dependency Cost

A task type that causes many dependencies is another potential parallelization bottleneck. Thus, based on the number of dependencies (sum of incoming and outgoing dependencies), we define a metric called dependency cost of a task type. Dependency cost of some task is proportional to the maximal number of dependencies caused by some instance of that task. Therefore, if task i has instances whose numbers of dependencies are in the array D_i, the dependencies cost of task i is:

$$d_i = \max(z), z \in D_i \tag{4}$$

Furthermore, using a control parameter p, we define the normalized dependency cost of task i as:

$$\overline{d_i(p)} = \frac{(d_i)^p}{\sum\limits_{j=1}^{N} (d_j)^p}, \quad 0 \leq p \leq \infty \tag{5}$$

Metric 3: Task Concurrency Cost

A task type that has low concurrency is another potential parallelization bottleneck. Concurrency of some instance is determined by the overall utilization of the machine during the execution of that instance. Thus, we define concurrency cost of some task to be inversely proportional to the average number of cores that are efficiently utilized during the execution of that task. Therefore, if task i has task instances which run for time $T_{i,j}$ while there are j cores efficiently utilized, the concurrency cost of task i is:

$$c_i = \frac{\sum\limits_{j=1}^{cores} \frac{T_{i,j}}{j}}{\sum\limits_{j=1}^{cores} T_{i,j}} \tag{6}$$

Again, using a control parameter p, we define the normalized concurrency cost of task i as:

$$\overline{c_i(p)} = \frac{(c_i)^p}{\displaystyle\sum_{j=1}^{N}(c_j)^p}, \quad 0 \le p \le \infty \tag{7}$$

Control Parameter p

Introduction of the parameter p provides the mechanism for controlling the mutual distance of the normalized costs for different tasks. For instance, let us assume that the application consists of two task instances, A and B, where A is two times longer than B. If the control parameter p_l is equal to 1, the normalized length costs for tasks A and B are 0.67 and 0.33, respectively. However, if the control parameter p_l is equal to 2, the costs for tasks A and B become 0.8 and 0.2, respectively.

Therefore, by changing parameter p of some metric, we can control the impact of that metric on the overall cost. For example, if the control parameter for length cost is 0, all task types will have the same normalized length cost, independent of the length of task instances. Thus, the length of tasks would have no impact on the overall cost. On the other hand, if the control parameter for length cost is infinite, the task type with the longest instance will have the normalized length cost of 1, while all other task types will have the normalized length cost of 0. This way, the impact of the task length on the overall cost would be maximized.

4.1.2 Heuristic 2: When to Stop Refining the Decomposition

The algorithm also needs a condition to stop the iterative search. Iterative search leads to fine grain decompositions that instantiate a very high number of tasks. An excessive number of tasks causes a very complex and computation intensive evaluation of the potential parallelism. Thus, to make the complete automatic search viable, we must adopt the exit condition that will prevent processing unnecessary decompositions.

To construct the Heuristic 2, we must create a system for rating the quality of a decomposition. Our basic rating system consists of two rules. First, out of all tested decompositions, the optimal decomposition is the one that achieves the highest parallelism. Second, if the optimal decomposition achieves the parallelism of s_{opt} and instantiates t_{opt} tasks, and some other decomposition i achieves the parallelism of s_i and instantiates t_i tasks, the relative quality of decomposition i compared to the optimal decomposition is:

$$Quality_i = \left(\frac{s_i}{s_{opt}}\right) \cdot \left(\frac{t_{opt}}{t_i}\right)^{exp\_tasks}, \quad 0 \le exp\_tasks \le 1 \tag{8}$$

Thus, the relative quality of some decomposition drops as the achieved parallelism drops and as the number of instantiated tasks increases. Furthermore, the parameter *exp_tasks* serves to tune the impact of the number of instantiated tasks.

Finally, Heuristic 2 mandates that the iterative search stops if the current decomposition has relative quality lower than some threshold value:

$$Quality_i < (Q_{threshold})^{\frac{cores}{S_{opt}}}, \quad 0 \le Q_{threshold} \le 1 \tag{9}$$

The right side of this expression increases with the increase of the parallelism of the optimal task decomposition. Thus, if the optimal found parallelism is close to the theoretical maximum (number of cores in the target machine), finding a better decomposition is unlikely, so the algorithm should tolerate only low quality degradations. On the other hand, if the optimal found parallelism is far from the theoretical maximum, the algorithm should be more aggressive in finding a better decomposition, and therefore allow high degradations of quality.

4.2   Tareador Environment for Automatic Exploration of Parallelism

In order to adapt Tareador for automatic exploration of parallelism, into the original environment we additionally introduced **Paramedir** and search **Driver**. Paramedir [6] (the non-graphical user interface to the Paraver) extracts parallelization metrics described in Sect. 4.1. On the other hand, the Driver iteratively explores parallelism by specifying in each iteration a different *list of tasks* that compose the current task decomposition (Fig. 8). More specifically, the Driver guides the environment through the following steps:

1. **Generate execution logs**: dynamically instrument the sequential application and derive execution logs.
2. **Select the starting decomposition**: put the whole main into one task.
3. **Estimate the parallelism of the current decomposition**: generate traces that estimate the parallelism of the current decomposition.
4. **If the exit condition is fulfilled, finish**: if the *Quality* of the current decomposition is unsatisfactory (Heuristic 2), end the search.
5. **Else, identify the parallelization bottleneck**: process the traces with Paramedir to derive metrics that identify the bottleneck task (Heuristic 1).
6. **Refine the current decomposition to increase parallelism**: break the bottleneck task into its children tasks, if any. Update the *list of tasks* that should be included in the next decomposition.
7. **Proceed to the next iteration**: go to step 3.

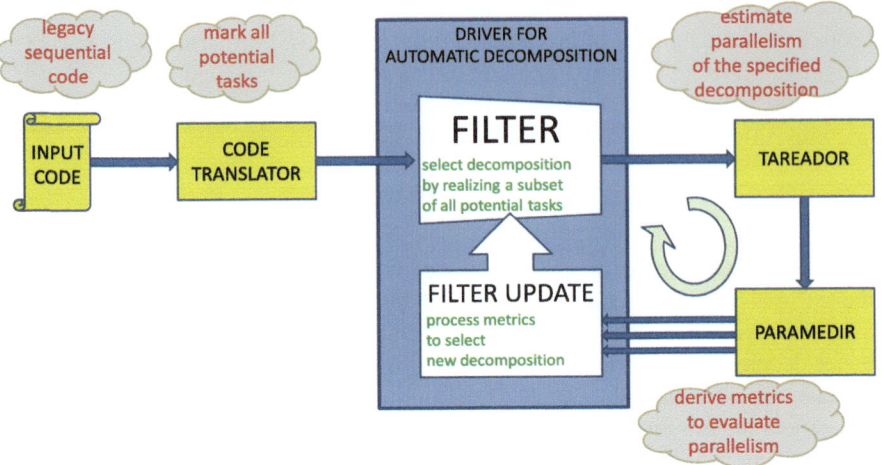

Fig. 8 Environment to automatically explore possible task decompositions

Table 1 Empirically
identified parameters
of the automatic search

p_l	p_d	p_c	exp_{tasks}	$Q_{threshold}$
1	1	3	$log_{10}1.5$	0.75

4.3 Experiments

Our experiments explore possible parallelization strategies for two well-known applications (Cholesky and LU factorization). We select a homogeneous multi-core processor as the simulated target platform. The goal of our experiments is to show that the proposed search algorithm, metrics and heuristics can find decompositions that provide significant parallelism.

Table 1 lists the empirically identified values for the parameters defined in Sect. 4.1. As already mentioned, the total cost function is a sum of length, dependency and concurrency cost (Eq. 1). Moreover, since our initial experiments showed that concurrency criterion prevails very rarely, we decided to increase the weight of the concurrency cost. Furthermore, in Eq. 8, we set the parameter exp_{tasks} so that the increase of task instances by a factor of 10 is equivalent to the decrease of parallelism by a factor of 1.5. Finally, in Eq. 9, parameter $Q_{threshold}$ was set empirically to allow sufficient quality degradation for a flexible search.

4.3.1 Illustration of the Iterative Search

To illustrate the algorithm we will use the example of parallelizing Cholesky sequential code on a simulated machine with 4 cores. Figure 9 presents (on the left) the code of Cholesky and illustrates (one the right) how the code can be encapsulated into tasks for various decompositions (*D*1–*D*6). Note that marked task types (boxes with numbers) may generate multiple task instances, and that

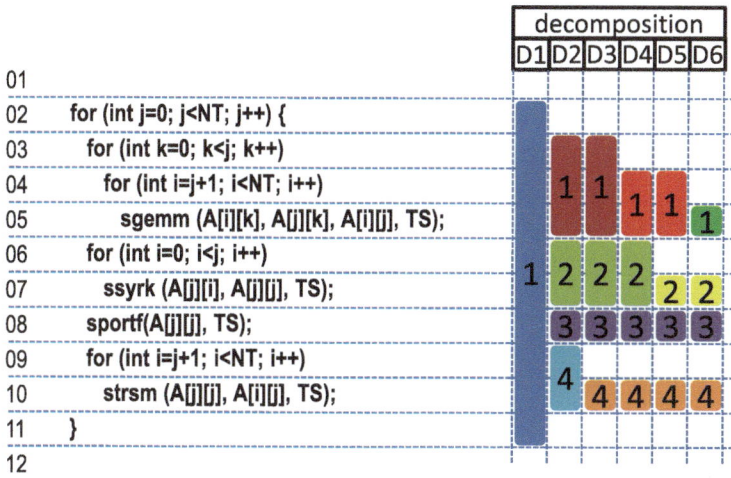

```
01
02        for (int j=0; j<NT; j++) {
03            for (int k=0; k<j; k++)
04                for (int i=j+1; i<NT; i++)
05                    sgemm (A[i][k], A[j][k], A[i][j], TS);
06            for (int i=0; i<j; i++)
07                ssyrk (A[j][i], A[j][j], TS);
08            sportf(A[j][j], TS);
09            for (int i=j+1; i<NT; i++)
10                strsm (A[j][j], A[i][j], TS);
11        }
12
```

Fig. 9 Cholesky: decomposition of the code into tasks

Table 2 Cholesky: task costs (Heuristic 1)

Decomposition	Speedup	Task #1				Task #2				Task #3				Task #4			
		$\bar{l}_i(1)$	$\bar{d}_i(1)$	$\bar{c}_i(3)$	$\bar{t}_i$	$\bar{l}_i(1)$	$\bar{d}_i(1)$	$\bar{c}_i(3)$	$\bar{t}_i$	$\bar{l}_i(1)$	$\bar{d}_i(1)$	$\bar{c}_i(3)$	$\bar{t}_i$	$\bar{l}_i(1)$	$\bar{d}_i(1)$	$\bar{c}_i(3)$	$\bar{t}_i$
D1	1.00	1.00	1.00	1.00	3.00												
D2	1.30	0.51	0.21	0.24	0.96	0.29	0.25	0.12	0.67	0.03	0.13	0.15	0.31	0.17	0.41	0.49	1.06
D3	1.49	0.59	0.44	0.48	1.51	0.34	0.18	0.22	0.74	0.04	0.18	0.27	0.48	0.03	0.21	0.03	0.27
D4	2.30	0.42	0.25	0.04	0.71	0.49	0.24	0.50	1.22	0.05	0.24	0.42	0.70	0.04	0.28	0.04	0.36
D5	3.41	0.72	0.27	0.11	1.10	0.12	0.17	0.13	0.43	0.09	0.25	0.61	0.95	0.07	0.30	0.15	0.52
D6	3.64	0.31	0.21	0.13	0.65	0.30	0.17	0.22	0.70	0.22	0.25	0.50	0.97	0.17	0.36	0.14	0.68

the code outside of marked tasks belongs to the *master task* (sequential part of execution that spawns worker tasks). Table 2 shows the speedup achieved in each decomposition and the costs that guide the iterative search. The algorithm starts from the most coarse-grain decomposition $D1$ that puts the whole execution into one task. There is only one task (#1, lines 2–11), which is automatically the critical task that needs to be broken. Refining $D1$ generates decomposition $D2$ that achieves the speedup of 1.30 (Table 2) and consists of 4 different task types (Fig. 9): #1 that covers the first loop (lines 3–5); #2 that covers the second loop (lines 6–7); #3 that covers function *spotrf_tile* (line 8); and #4 that covers the third loop (lines 9–10). Heuristic 1 identifies task #4 as the most critical, mostly due to its high concurrency cost. Thus, the following decomposition ($D3$) breaks the task #4 and obtains the parallelism of 1.49. In $D3$, the algorithm identifies task #1 as the bottleneck (due to its high length). Further iterations of the algorithm pass through decompositions $D4$, $D5$ and $D6$ that provide speedups of 2.30, 3.41 and 3.64, respectively.

4.3.2 Results

This subsection presents the results obtained by applying our algorithm on a set of applications. For each application, we present four plots that illustrate the process

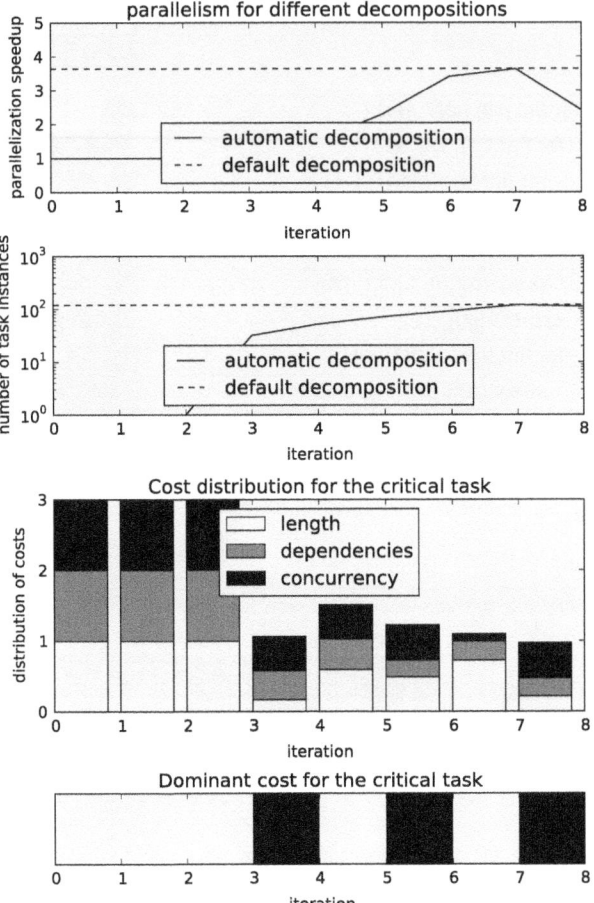

Fig. 10 Cholesky on 4 cores

of automatic task decomposition. The first plot presents the parallelism of all tested decompositions – the speedup over the sequential execution of the application. The second plot shows the number of task instances generated by each decomposition. Also, the first two plots show the parallelism and the number of instances in the *reference task decomposition* (the decomposition selected and implemented by an expert programmer). The third plot presents the cost distribution for the bottleneck task of each iteration. Finally, the fourth plot shows the most dominant cost for the bottleneck task.

The proposed search algorithm finds decompositions with very high parallelism, sometimes finding the decomposition manually selected by an expert programmer. The algorithm finds the reference decomposition for Cholesky in iteration 7 (Fig. 10). In order to get to this decomposition, the algorithm refines decompositions based on the concurrency criterion in iterations 3 and 5. Soon after finding the

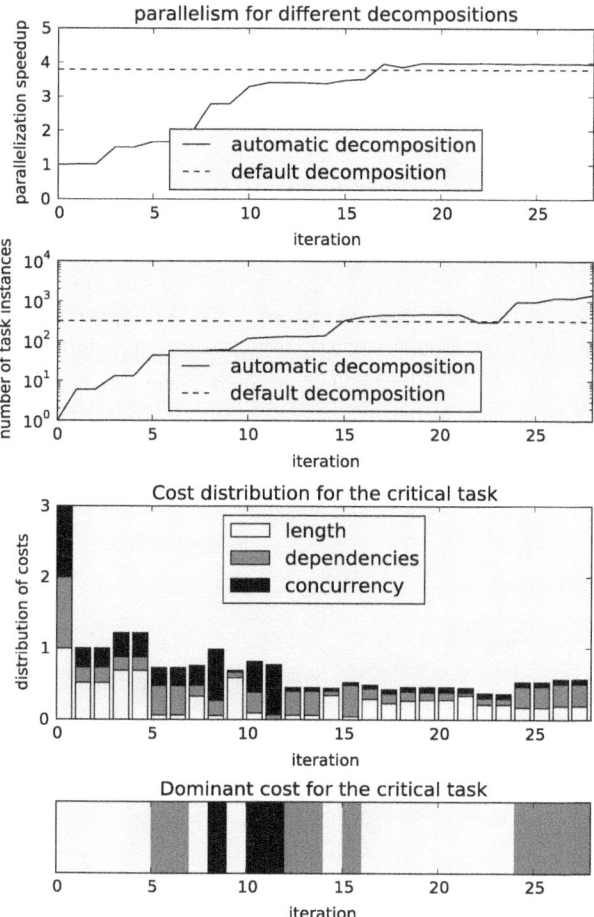

Fig. 11 Sparse LU on 4 cores

reference decomposition, the algorithm passes through the decomposition that activates the mechanism for stopping the search (Heuristic 2).

Sparse LU (Fig. 11), as a more complex application, demonstrates the power of our search. Compared to Cholesky, Sparse LU forces the algorithm to use various bottleneck criteria through the exploration of decompositions. It is interesting to note that the search finds a wide range of decompositions (iterations 17–28) that provide higher parallelism than the reference decomposition. In this case, it is unclear which of these decompositions is the optimal one. Quantitative reasoning suggests that the optimal task decomposition is the one that provides highest parallelism with the lowest number of created task instances. Following this reasoning, the optimal decomposition (iteration 22) achieves the speedup of 3.98 with the cost of 301 instantiated tasks (note the sudden drop in the number of task instances). On the

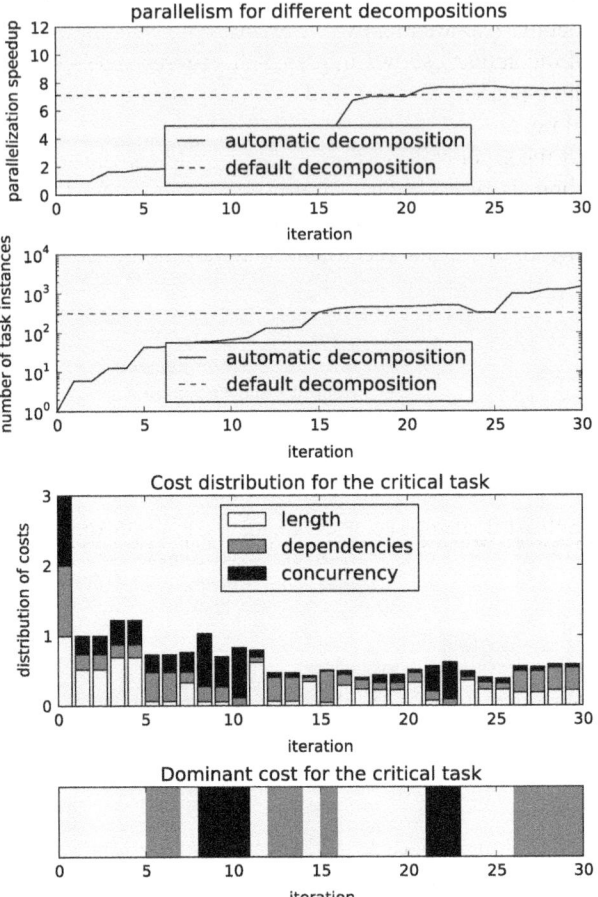

Fig. 12 Sparse LU on 8 cores

other hand, qualitative reasoning suggests that, within a set of decompositions that provide similar parallelism generating a similar number of instances, the optimal decomposition is the one that is the easiest to express using semantics offered by the target parallel programming model. For example, our algorithm may find a decomposition that extracts very irregular parallelism that cannot be expressed using a fork-join programming model. In that case, it is programmer's responsibility to, out of few offered efficient task decompositions, identify the one that can be straightforwardly implemented using a specific programming model.

It is also interesting to study how the algorithm adapts to the target parallel machine. Changing the parallelism of the target machine changes the simulation of the parallel execution of the tested decomposition. Thus, changes the normalized concurrency cost, while dependency and length cost remain the same. Figures 12 and 13 illustrate potential decompositions for Sparse LU for executing on machines

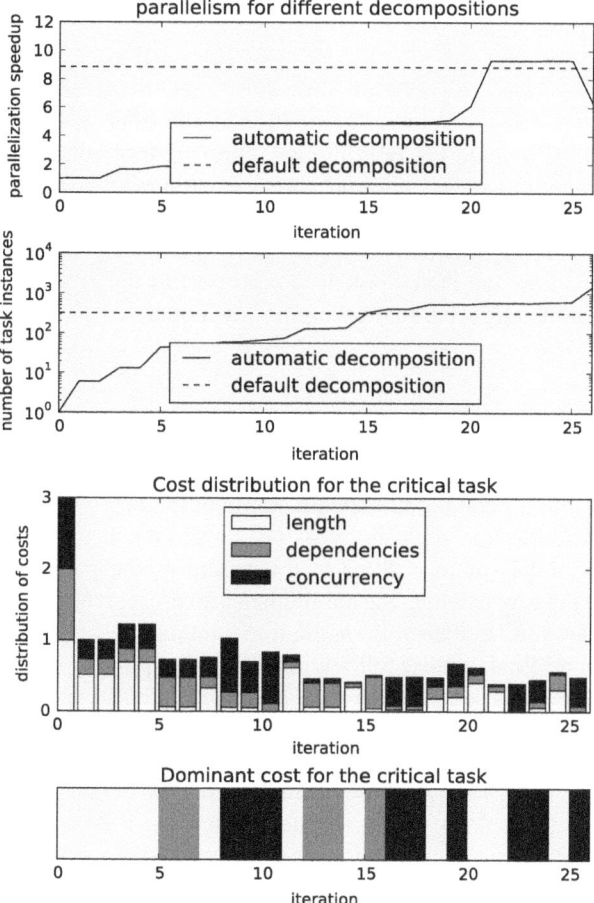

Fig. 13 Sparse LU on 16 cores

with 8 and 16 cores. In the experiments with 8-core target machine (Fig. 12), the reference decomposition achieves the speedup of 7.1 at the cost of generating 316 task instances. The automatic search finds a wide range of decompositions (iterations 21–30) that provide slightly higher parallelism than the reference decomposition. On the other hand, in the experiments with 16-core target machine (Fig. 13), the reference decomposition achieves the speedup of 8.85 (316 instances). The algorithm finds only five decompositions (iterations 21–25) that provide higher parallelism than the default decomposition. It is also interesting to note that in the experiment with 16-core target machine, the algorithm more often refines the decomposition using the concurrency criterion. This happens because, despite the fine granularity of decompositions, the algorithm cannot find decomposition with parallelism close to the theoretical maximum of 16 (number of cores in the target machine).

5 Discussion: Tareador in the Big Scheme of Things

This section describes our idea of Tareador's role in the complete solution for parallelizing applications. As already mentioned, the solution for parallelization of applications must include not only the development tool but also the appropriate parallel programming model and the parallelization workflow. Parallel programming model should be chosen so the results obtained by the development tool can be useful in expressing parallelism. On the other hand, the parallelization workflow should describe how the development tool is used in the process of parallelizing applications for the target parallel programming model.

5.1 OmpSs (OpenMP 4.0)

In order to design a good parallelization solution, the target parallel programming model should follow the philosophy of the development tool. Different parallel programming models mainly differ in the nature of the parallelism that can be exploited, and the way in which the parallelism is expressed. Most of the mainstream parallel programming models rely on the fork-join parallelism [7, 8] Fork-join parallelism in the original context follows all-or-nothing philosophy – it distinguishes between the sequential sections that have no parallelism and parallel sections where everything is parallel. Thus, these programming model allow expressing which code sections are sequential and which parallel. Other programming models rely on the dataflow paradigm [9, 10]. Dataflow parallelism tries to define for each pair of tasks whether they are dependent or not. In most of the dataflow programming models, the programmer expresses memory usage of each task type and then the runtime system dynamically calculates intertask dependencies.

In a parallelization solution that includes Tareador, we propose using OmpSs as the target parallel programming model. OmpSs [9] is a programming model developed in Barcelona Supercomputing Center as a forefront for OpenMP [7]. In fact, most of the main OmpSs ideas are already introduced in OpenMP 4.0 standard [11]. OmpSs is a directive-based dataflow parallel programming model. Figure 14 illustrates how OmpSs extracts parallelism in Cholesky kernel. Figure 14a shows the sequential Cholesky code (code in black-bold) and the OmpSs pragma directives needed to expose parallelism (code in gray). Thus, if a programmer wants some function to execute as a task, she must annotate the function declaration with the pragma directive that specifies the directionality of each function argument (input, output, inout). Based on these directives, the runtime system dynamically generates the dependency graph of all task instances (Fig. 14b). Once the dependency graph is generated, the runtime can execute task instances out of order, as long as dependencies are satisfied. This type of dataflow execution allows extracting very irregular parallelism that cannot be expressed with fork-join parallelism.

a

```
01    #pragma omp task inout ([TS][TS]A)
02    void sportf (float *A, int TS);
03    #pragma omp task input ([TS][TS]T) inout ([TS][TS]B)
04    void strsm (float *T, float *B, int TS);
05    #pragma omp task input ([TS][TS]A, [TS][TS]B) inout ([TS][TS]C)
06    void sgemm (float *A, float *B, float *C, int TS);
07    #pragma omp task input ([TS][TS]A) inout ([TS][TS]C)
08    void ssyrk (float *A, float *C, int TS);
09
10    void cholesky (int NT, float *A[NT][NT]) {
11      for (int k=0; k<NT; k++) {
12        ◯ sportf(A[k][k], TS);
13        for (int i=k+1; i<NT; i++)
14          ⬤ strsm (A[k][k], A[k][i], TS);
15        for (int i=k+1; i<NT; i++) {
16          for (int j=k+1; j<i; j++)
17            ⬤ sgemm (A[k][i], A[k][j], A[j][i], TS);
18          ⬤ ssyrk (A[k][i], A[i][i], TS);
19        }
20      }
21    }
```

b

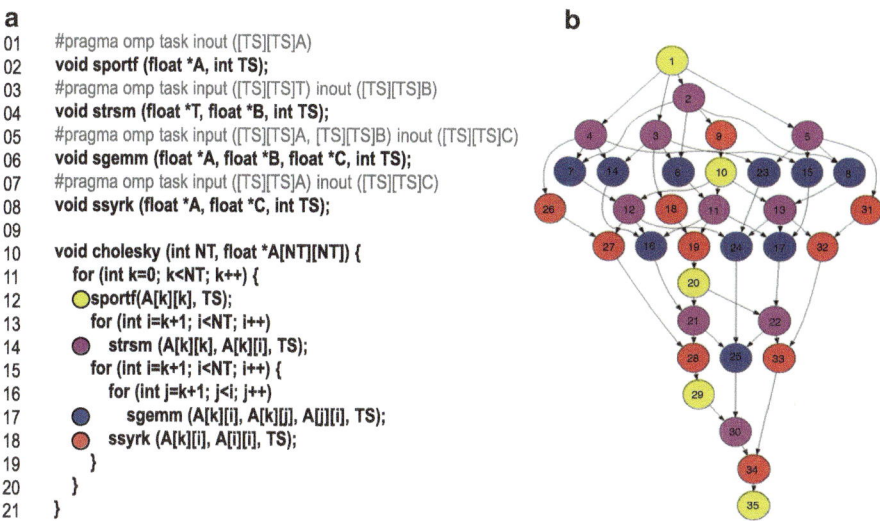

Fig. 14 Cholesky parallelized with OmpSs. (**a**) Source code. (**b**) Task dependency graph (**Note**: The code listing marks for each function the color of the node that represents it in the dependency graph)

Furthermore, numerous research papers prove that OmpSs outperforms OpenMP in many well-known scientific applications [12, 13].

We consider OmpSs programming model as a specially good fit for Tareador development tool. For a specified task decomposition, Tareador dynamically calculates the memory usage of each task and then calculates potential inter-task dependencies. Therefore, the information obtained by Tareador naturally maps to the information needed to express OmpSs parallelism (directionality of function arguments inside pragmas). Furthermore, OmpSs facilitates automatic generation of parallel code. Being a directive-based parallel programming model, OmpSs adds pragma annotations, but maintains the original code structure of the sequential application. Therefore, it is easy to suggest how to change the sequential code in order to parallelize it, much easier than in the case when the code structure needs to be modified (for instance, in the case of pthreads [14]).

5.2 Parallelization Workflow

Figure 15 illustrates the parallelization workflow for using Tareador. At the current stage of development, Tareador serves as a tool to explore potential parallelism in sequential applications. Thus, starting from the sequential code (#0), Tareador LLVM module (#1) generates execution logs (#2). Processing the logs, Tareador GUI explores various task decompositions (#3) and evaluates their potential

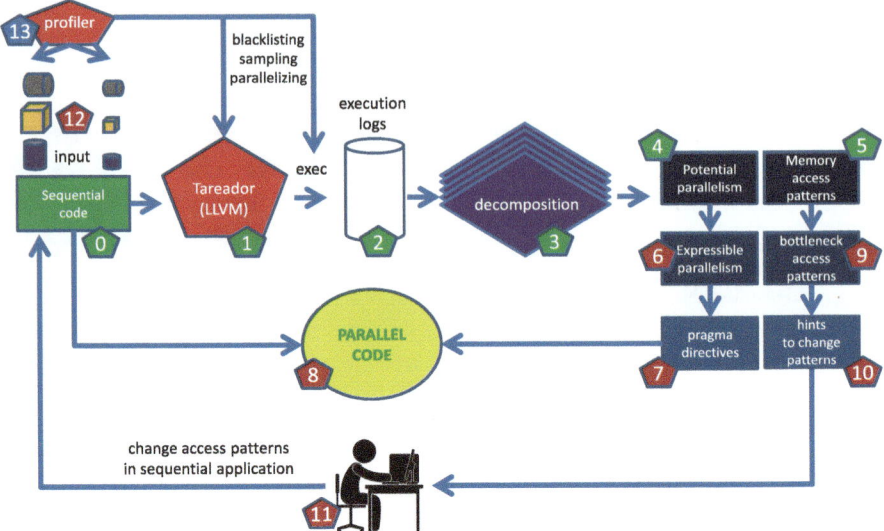

Fig. 15 Parallelization workflow using Tareador

parallelism (#4). Furthermore, Tareador collects and visualizes memory access patterns of all all tasks (#5). This part of development is already finished.

The rest of the Figure represents planed future development of Tareador environment. Currently, Tareador evaluates only unbounded parallelism – parallelism not limited by constraints of the target parallel programming model. In other words, Tareador identifies two tasks as dependent iff the first task writes at least on byte that the second task reads. However, Tareador omits to consider whether the tested decomposition can be implemented – whether the targeted parallel programming model offers parallelization primitives that can express the identified parallelism. Thus, the next step for Tareador is to evaluate the portion of the potential parallelism that can be expressed using the target parallel programming model (#6). Once this step is finished, Tareador should output the content of pragma primitives that are required to expose parallelism (#7). Thus, the correct parallel code could be obtained (#8). However, in most applications, the parallel efficiency of the code obtained this way will be unsatisfactory.

The potential of automatic parallelization is limited by unfavorable access patterns to some memory objects. If simple decomposition of the sequential code into tasks cannot provide sufficient expressible parallelism, Tareador should pinpoint the memory objects with unfavorable access patterns (#9). Also, Tareador should suggest to the programmer how to change these access patterns (#10) in order for the automatic parallelization to be more efficient. Then, the programmer should manually change the culprit access patterns (#11) and pass the new application for the next attempt of automatic parallelization. It is important to note that in the proposed workflow, the programmer modifies only the sequential application and adapts it so the automatic parallelization could be more efficient. Finally, Tareador should process the application for different inputs (#12) that exercise potentially different parts of the source code. The parallelization strategy that suits all tested inputs should be accepted as the final one.

Nevertheless, Tareador development will continuously dedicate to optimization of the instrumentation (#13). Currently we are developing the profiler tools whose output should facilitate optimization of the original Tareador instrumentation. Firstly, the profile information should allow better blacklisting of code sections that are promoted into potential tasks. Furthermore, profiler should allow sampling by pinpointing smaller parts of execution that can be representatives of the whole run. Also, profile information should suggest how to parallelize the instrumentation by separately instrumenting different parts of execution and then merging the obtained logs. Finally, the profiler should monitor the process of reducing the problem size while preserving the characteristic behavior of the application.

6   Related Work

Numerous tools to assist parallelization have been proposed in the past years both from the academia and the industry. Regarding tools proposed by the academia, the ones closest to the environment that has been proposed in this paper are Embla, Kremlin, and Alchemist. In particular, Embla [15] is a Valgrind-based tool that estimates the potential speed-up for Cilk programs. On the other hand, Kremlin [16] identifies regions of a serial program that can be parallelized with OpenMP and proposes a parallelization planner for the user to parallelize the target program. Finally, Alchemist [17] identifies parts of code that are suitable for thread-level speculation. The major drawbacks of these tools are that they are limited to fork-join parallelism and that they offer very little qualitative information about the target program (no useful visualization support).

On the other side, the industry have also been recently developing their solutions for assisted parallelization. For example, Intel's Parallel Advisor [18] assists parallelization with Thread Building Blocks (TBB) [19]. Parallel Advisor provides timing profile that suggests to the programmer which loops should be parallelized. Critical Blue provides Prism [20], a tool to do "what-if" analysis that anticipates the potential benefits of parallelizing certain parts of the code. Vector Fabrics provides Pareon [21], another tool for "what-if" analysis to estimate the benefits of parallelizing loop iterations. All the three mentioned tools provide rich GUI and visualization of the potential parallelization. However, none of the tools offers automatic exploration of parallelization strategies. Moreover, they do not provide any API to automate the search for the optimal parallelization strategy as the one proposed in this paper.

7 Conclusion and Future work

The software community is facing a paramount task of parallelizing the existing body of sequential applications. Current mainstream hardware provides extremely high parallelism, but state-of-the-art sequential software cannot take advantage of this abundance of resources. To adapt current applications for the novel hardware, we must approach the challenge of parallelization.

In order to tackle this issue, in this paper we present Tareador – a tool for assisted parallelization. Tareador allows to the programmer to easily browse various parallelization strategies and choose the one that promises the highest parallelization potential. Furthermore, Tareador provides very rich visualization of the results, offering deeper insight into the potential parallelism and pinpointing the culprits for low performance. We showed how Tareador is used for teaching parallelism at the University courses, as well as for actually parallelizing sequential applications.

At the current stage of development, Tareador is useful for exploring the potential parallelism inherent in the applications. However, we describe our future development directions in order to upgrade Tareador into a tool for automatic parallelization of sequential applications. We also blueprint the parallelization process that the programmer should follow in order to use Tareador to port sequential application to OmpSs parallel programming model.

References

1. Lattner, C., Adve, V.: LLVM: a compilation framework for lifelong program analysis and transformation, San Jose, pp. 75–88 (2004)
2. Girona, S., Labarta, J., Badia, R.: Validation of dimemas communication model for MPI collective operations. In: EuroPVM/MPI'2000, Lake Balaton (2000)
3. Pillet, V., Labarta, J., Cortes, T., Girona, S.: PARAVER: a tool to visualize and analyze parallel code. In: WoTUG-18, Manchester (1995)
4. Gansner, E.R., North, S.C.: An open graph visualization system and its applications to software engineering. Software – Practice and Experience **30**(11), 1203–1233 (2000)
5. Subotic, V., Ferrer, R., Sancho, J.C., Labarta, J., Valero, M.: Quantifying the potential task-based dataflow parallelism in MPI applications. In: Euro-Par (1), Bordeaux, pp. 39–51 (2011)
6. Jost, G., Labarta, J., Gimenez, J.: Paramedir: a tool for programmable performance analysis. In: International Conference on Computational Science, Kraków, pp. 466–469 (2004)
7. Dagum, L., Menon, R.: OpenMP: an industry-standard API for shared-memory programming. Comput. Sci. Eng. **5**, 46–55 (1998)
8. Blumofe, R.D., Joerg, C.F., Kuszmaul, B.C., Leiserson, C.E., Randall, K.H., Zhou, Y.: Cilk: an efficient multithreaded runtime system. J. Parallel Distrib. Comput. **37**, 55–69 (1996)
9. Duran, A., Ayguadé, E., Badia, R.M., Labarta, J., Martinell, L., Martorell, X., Planas, J.: Ompss: a proposal for programming heterogeneous multi-core architectures. Parallel Process. Lett. **21**(2), 173–193 (2011)
10. K. Fatahalian, Horn, D.R., Knight, T.J., Leem, L., Houston, M., Park, J.Y., Erez, M., Ren, M., Aiken, A., Dally, W.J., Hanrahan, P.: Memory – sequoia: programming the memory hierarchy. In: SC, New York, p. 83 (2006)

11. OpenMP Architecture Review Board: OpenMP Application Program Interface Version 4.0. http://www.openmp.org/mp-documents/OpenMP4.0.0.pdf. Active on July 2013
12. Pérez, J.M., Badia, R.M., Labarta, J.: A dependency-aware task-based programming environment for multi-core architectures. In: CLUSTER, Tsukuba, pp. 142–151 (2008)
13. Marjanovic, V., Labarta, J., Ayguadé, E., Valero, M.: Overlapping communication and computation by using a hybrid MPI/SMPSs approach. In: ICS, Tsukuba, pp. 5–16 (2010)
14. Nichols, B., Buttlar, D., Farrell, J.P.: Pthreads Programming. O'Reilly & Associates, Sebastopol (1996)
15. Mak, J., Faxén, K.-F., Janson, S., Mycroft, A.: Estimating and exploiting potential parallelism by source-level dependence profiling. In: Euro-Par (1), Ischia, pp. 26–37 (2010)
16. Garcia, S., Jeon, D., Louie, C.M., Taylor, M.B.: Kremlin: rethinking and rebooting gprof for the multicore age. In: PLDI, San Jose, pp. 458–469 (2011)
17. Zhang, X., Navabi, A., Jagannathan, S.: Alchemist: a transparent dependence distance profiling infrastructure. In: CGO '09, Seattle (2009)
18. Intel Corporation: Intel Parallel Advisor. http://software.intel.com/en-us/intel-advisor-xe. Active on 10.11.2014
19. Pheatt, C.: Intel threading building blocks. J. Comput. Sci. Coll. **23**, 298–298 (2008)
20. Critical Blue: Prism. http://www.criticalblue.com/. Active on 10.11.2014
21. Vector Fabrics: Pareon. http://www.vectorfabrics.com/products. Active on 10.11.2014

Tuning Plugin Development for the Periscope Tuning Framework

Isaías A. Comprés Ureña and Michael Gerndt

Abstract Periscope, the automatic performance analysis tool, was extended in the European AutoTune project to support automatic tuning. As part of the extension, the tool provides a framework for the development of automatic tuners. The Periscope Tuning Framework (PTF) facilitates the development of advanced tuning plugins by providing the Tuning Plugin Interface (TPI). The tuners are implemented as plugins that are loaded at runtime. These can access the performance analysis features of Periscope as well as its automatic experiment execution support. The partners in AutoTune developed tuning plugins for compiler flag selection, MPI library parameters, MPI IO, master/worker applications, parallel pattern applications, and energy efficiency. This presentation will outline the development of tuning plugins and gives examples from the plugins developed in AutoTune.

1 Introduction

Current computing hardware is undeniably parallel. In mobile and embedded systems, low power multi-core CPUs are already the norm. In high-end workstations, common configurations consist of multi- and many-core CPUs with accelerators that integrate thousands of vector processing units. In HPC systems, the amount of parallel processing capacity is further increased by attaching nodes with already highly parallel processing hardware through high performance networks. Tools that help developers extract acceptable performance on these parallel systems are of increased importance given the current trend towards parallelism, and this is more so in the context of HPC, where the trend can be observed in its strongest form.

The research leading to these results has received funding from the European Union Seventh Framework Programme (FP7/2007-2013) under grant agreement number 288038 (AutoTune Project, www.autotune-project.eu).

I.A. Comprés Ureña • M. Gerndt (✉)
Institute of Informatics, Technical University of Munich (TUM), Boltzmannstr. 3,
85748 Garching, Germany
e-mail: compresu@in.tum.de; gerndt@in.tum.de

© Springer International Publishing Switzerland 2015
C. Niethammer et al. (eds.), *Tools for High Performance Computing 2014*,
DOI 10.1007/978-3-319-16012-2_5

Current tools used in HPC systems are actively improved in terms of their scalability. In this area of computing, tools that are not able to adapt to thousands of MPI processes in the global scope and tens of threads at the node level, risk limiting their usability and, in the worst case, becoming obsolete.

The improved scalability of the tools to cope with the increased amount of parallelism is today not sufficient. In addition to the tools actually running efficiently in these new systems, the presentation of information to the user needs to be adapted, and aspects of performance that were addressed directly by the users now should, as much as possible, be automated.

Most of the current tools focus on the presentation of information for the users, while a few focus on the automation of the optimization process. The tool presented in this paper focuses on the later. The Periscope Tuning Framework (PTF) provides a generic platform that can be used by programmers to develop automatic tuners. The framework is also distributed with automatic tuners that can be used directly by users. PTF differs from other tools, in that the tuning process is based on a mix of expert knowledge, Periscope analyses and search strategies.

The aim of this paper is to first introduce the Periscope Tuning Framework (PTF) from an abstract software architecture, and second give a concrete introduction to the actual C++ framework with a sample implementation. Section 2 introduces the general architecture of the framework. In Sect. 3 the expert knowledge utilized in the implementation of the distributed tuning plugins is explained. The interface provided for the development of automatic tuners is introduced in Sect. 4, followed by an explanation on how to use it in Sect. 5. After that, a sample OpenMP plugin that optimizes thread counts is presented in Sect. 5.3. The Sects. 5.4 through 5.8 cover some of the features of PTF, as individual extensions to the OpenMP plugin. Finally, results from applying the sample OpenMP plugin to the NAS Parallel Benchmark's Block Tridiagonal solver are presented in Sect. 6.

1.1   Related Work

The HPC tools community has produced a rich collection of tools that scale to current petascale systems and continue to improve in anticipation of exascale systems. The community produces tools with relatively low overlap; the tools tend to focus on precise areas of interest depending on the type of parallel hardware platform, programming model, runtime environment and programming language. From the vast list of available tools, some of them will be briefly mentioned in this section.

Scalasca [4] provides information about possible performance bottlenecks; in particular, it tries to identify those related to communication and synchronization. Vampir [3] focuses on the collection and graphical representation of performance data, in particular traces and profiles. The TAU Performance System [8] is a toolkit for tracing and performance analysis of parallel applications. A common infrastructure for the collection of traces and generation of profile information is ScoreP [7].

Tools that focus particularly on automatic tuning have been widely developed by compiler and runtime vendors. Active Harmony [9] has targeted the task of automating tuning with a more general scope, and has been successful in optimizing both parallel and single node software automatically.

The Periscope Tuning Framework (PTF) and Active Harmony differ from other tools in that they attempt to optimize the application automatically. PTF differs from Active Harmony in the way that it attempts to reduce parameter spaces; in PTF, this is done through the use of algorithms based on expert knowledge together with Periscope analyses, while in Active Harmony this is achieved through parameter sensitivity tests and elaborate search mechanisms.

2   Periscope Tuning Framework (PTF)

The Periscope Tuning Framework (PTF) consists of Periscope and the tuning plugins developed in the AutoTune project. The most important novelty of PTF is the close integration of performance analysis and tuning. It enables the plugins to shrink the search space, to increase the efficiency of the tuning plugins and to gather detailed information during the evaluation of tuning scenarios. The performance analysis determines information about the execution of an application in the form of performance properties.

The overall architecture of the framework is shown in Fig. 1. It consists of the user interface, frontend, analysis agent network, and the MRI monitor that is linked to the application.

The analysis capabilities of Periscope are implemented by all these layers. The user interface allows to inspect the found performance properties in Eclipse, the frontend triggers performance analysis strategies that are executed by the analysis agents, and the MRI monitor measures performance data required for the automatic identification of performance properties.

The framework goes beyond automatic performance analysis and allows for the tuning of applications with respect to various performance aspects. It provides a rich collection of tools for implementing *tuning plugins*. Tuning plugins optimize specific aspects of applications' performance. Performance aspects can be hardware and software platform, programming model or parallel pattern specific. The plugins follow a predefined *tuning model* (Sect. 2.1) that defines the sequence of operations that all plugins have to implement. The operations are defined by the *Tuning Plugin Interface (TPI)* that is described in Sect. 4. Plugins are loaded dynamically and can be provided in source or binary form.

The sequence of TPI operations is determined by the frontend. It calls the plugin operations and implements the interface between the plugins and the rest of Periscope. The plugins typically investigate a number of variants of the execution to identify optimizations. In this process, analysis information is used to shrink the search space and to improve the search efficiency of plugins. On request of the

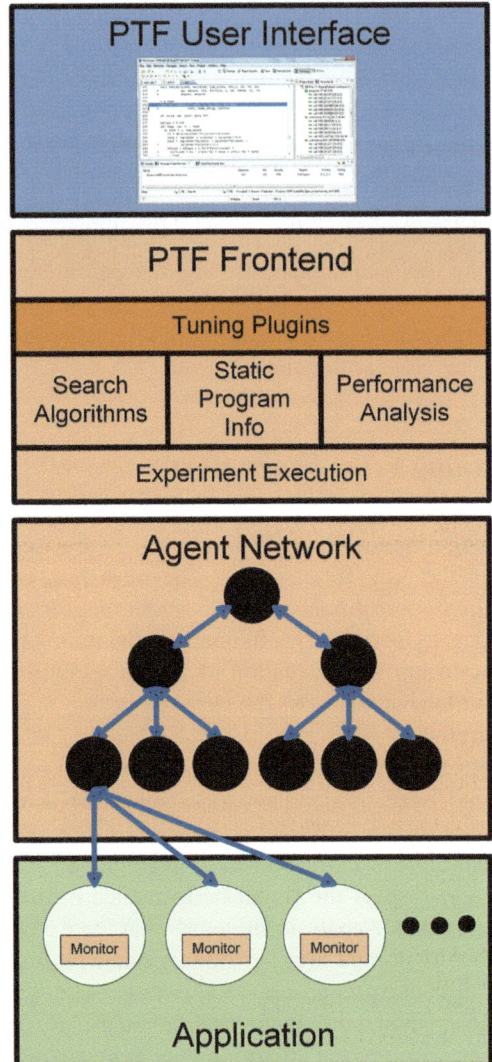

Fig. 1 Architecture of the Periscope Tuning Framework

plugins, analysis strategies are executed and the found properties are returned to the plugin to provide dynamic information.

In addition to dynamic information, Periscope also provides static information that is gathered during program instrumentation and passed to the system in the *Standard Intermediate Program Representation (SIR)*. It provides, for example, the available program regions which can be investigated for tuning purposes.

Tuning plugins run experiments and measure the effect of an optimization on the tuning objective. While the *tuning objectives* are standard Periscope properties, the implementation of an optimization is realized by *tuning actions*. These tuning actions might be executed at runtime to, for example, assign a certain value to a variable. They are propagated by the analysis agents to the MRI monitor, which performs the action at a predefined state of the program execution.

The implementation of tuning plugins is supported by a set of standard *search algorithms* that can be used to generate the variants of the tuning space. The search algorithms are dynamically loaded on request of tuning plugins. Thus, as the plugins itself, search algorithms can be provided in source or binary form. Alternatively, a plugin may also implement its own search algorithm internally.

2.1 Tuning Model

The tuning plugins of PTF follow the tuning model shown in Fig. 2. The plugin is first initialized and it creates the tuning parameters defining the tuning space. These tuning parameters either result from the tuning approach or might be given via a configuration file to the plugin. Then the plugin goes through one or more tuning steps. Each tuning step may start with a pre-analysis that runs one of Periscope's analysis strategies to gather properties of the application. Based on these properties, the variant space is determined, i.e., a subspace of the tuning space. The variant space and the variant context then determine the search space.

The plugin will use one of PTF's search algorithms or a plugin-specific search algorithm to search for the best variant in the space. It constructs for each of the variants a scenario. The framework provides single- and multi-step algorithms. A single-step algorithm generates all the different scenarios at the beginning while the other class goes through multiple search steps. In each search step, a multi-step algorithm will create new scenarios based on the results of the previous search step.

The scenarios are then executed on the target system and the objectives are measured. Before the execution of an experiment starts, the scenario can be prepared, e.g., in this step the application can be recompiled. While the experiment is executed, the plugin can request any PTF analysis strategy to be run during the experiment to get detailed information on the effect of the tuning.

Once all the search steps were executed, the plugin can decide to go for additional tuning steps, determining a new variant space based on the previous results or results of a new pre-analysis. When the last search step is completed, the plugin will generate the advice for the user.

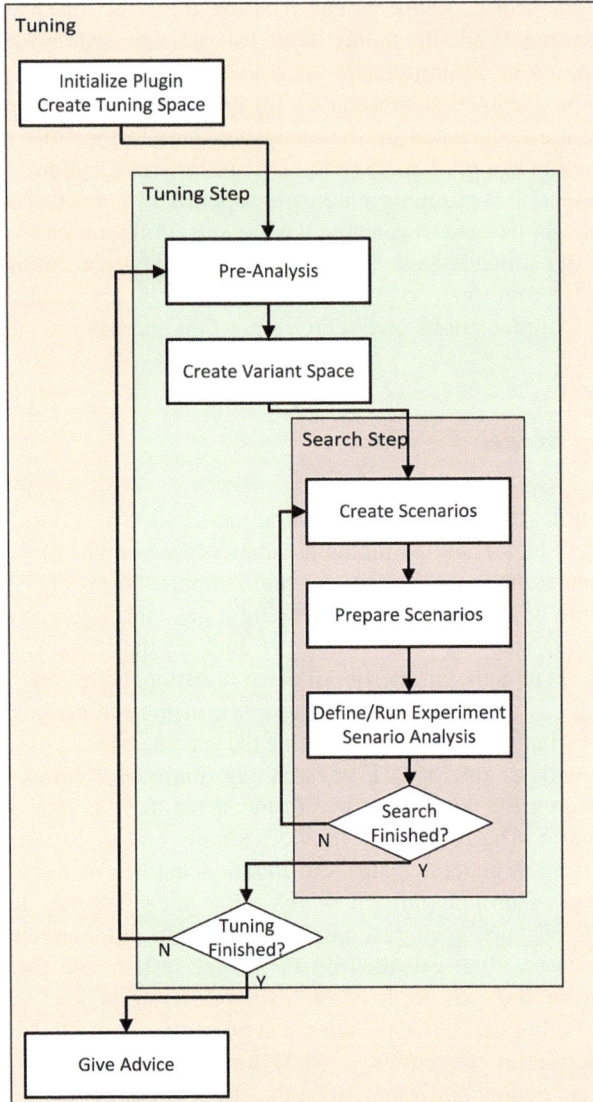

Fig. 2 Model of the tuning strategy of a plugin

3 Use of Expert Knowledge in Tuning Plugins

The tuning space of certain program aspects is typically quite large requiring a lot of experiments if it would be explored exhaustively. Therefore, tuning plugins in PTF explore expert knowledge about the tuning aspect in their tuning strategy. This section presents the different tuning plugins currently provided in PTF and how they explore expert knowledge in their tuning strategy.

3.1 MPI Parameters

The *MPI Parameters plugin* developed by Universitat Autònoma de Barcelona (UAB) explores tuning parameters of the MPI library. They provide parameters such as the eager limit, i.e., defining for which message sizes the library applies the eager transfer protocol, and the internal message buffer size. Libraries like Intel MPI, IBM MPI, and OpenMPI provide between 50 and 150 of such parameters where several of them are performance related.

Besides exhaustive search, the plugin offers a genetic search strategy that optimizes the parameter configuration based on an evolutionary approach. Instead of going through all the combinations, it goes through sequences of populations where new individual variants are generated based on the best configurations by crossover and mutation operations.

The plugin implements a special automatic strategy for tuning the eager limit. It is based on an analysis of the messages sizes transferred in the application. Based on the sizes, PTF identifies a property that indicates that the application is sensitive to changes of the eager limit. The plugin takes this property and the distribution of messages in the possible range for the eager limit to shrink the search space.

3.2 DVFS

The Leibniz Supercomputing Centre implemented the *DVFS plugin* that tunes the energy usage of applications based on setting the clock frequency of the processors. It tunes the global frequency setting for the entire application but also recommends special frequencies for individual program regions. For example, memory bound program regions can be executed with a smaller clock frequency then compute bound region without significantly increasing the execution time.

The plugin is based on an energy model that predicts the best frequency and execution time. To apply the model, the plugin first uses PTF's analysis to gather hardware counter metrics. Then, the prediction is performed. The predicted frequency and the next lower and higher frequency are experimentally evaluated.

For tuning specific program regions, a PTF analysis strategy is used to gather the hardware counter values and also determines regions that are suited for energy tuning. These regions have to have a granularity larger than a millisecond. When the experiments for the three frequency are executed, the plugin requests also the energy values for those suited regions. Based on these measurements, the plugin can give a recommendation which of the three frequencies is best suited for each of the regions.

3.3   Parallelism Capping

The *Parallelism Capping plugin* explores the selection of the best number of threads for OpenMP parallel regions to reduce the applications energy consumption. It explores the cross product of setting this parameter for all parallel regions. This huge search space is then evaluated by the Generalized Differential Evolution 3 (GDE3) algorithm [6] for multi-objective tuning. Multi-objective tuning is used here to find the best points on the Pareto curve for the possibly conflicting objective to reduce the energy and to reduce the execution time.

3.4   Master Worker

The *Master Worker plugin* developed by Universitat Autònoma de Barcelona (UAB) targets the tuning of the execution time of MPI applications following the master-worker parallel pattern. The master splits the computational domain into chunks and distributes the tasks to the workers. The tuning parameters are the block sizes for the chunks as well as the number of workers.

This plugin executes the tuning in two steps. First, its runs an analysis of the load imbalance between the workers based on a block sizes of one. From these analysis results, the plugin uses a performance model to predict the best block size. In the second step, the analysis is repeated with the new block size and the best number of workers is given by a second performance model.

3.5   Parallel Pattern

The *Parallel Pattern plugin*, developed at University of Vienna, optimizes the execution time of applications following a pipeline design. These applications are written in the high-level programming framework developed in the context of the European PEPPHER project [2]. Its tuning parameters are the replication factor for stages in the pipeline, the size of buffers to hold data packets between the stages, the number CPU cores and GPUs as well as the scheduling policy of the underlying runtime system.

The tuning strategy uses the results of an analysis of the application's execution, as well as user given hints, to shrink the search space. The analysis determines limiting stages in the pipeline. The tuning strategy then focuses on these limiting stages and investigates the stage replication factor and the buffer size settings. These parameters are evaluated in combination with the global tuning parameters (i.e., the CPUs, GPUs, and the scheduling policy).

3.6 MPI-IO

The *MPI-IO plugin*, developed at Technical University of Munich (TUM), improves the execution of MPI-IO operations in parallel applications. MPI-IO offers two tuning techniques for IO operations. Two-phase IO first gathers IO requests of the individual ranks in a certain number of aggregators. They aggregate the requests into larger requests that are then more efficiently executed by the IO subsystem. The other technique is called data sieving. The MPI ranks use a sieving buffer and fill this buffer with a data block of the file. Instead of writing or reading individual requests for the parts of the block, these requests read/write the data from/to the block.

As tuning parameters, MPI-IO considers the number of aggregators in two-phase IO (for collective operations) and the buffer size for aggregated requests. Additionally, the buffer size for the data-sieving optimization of individual MPI-IO operations.

Instead of exploring the cross product of these tuning parameters, the plugin applies two tuning steps. It exploits that the buffer sizes and the number of aggregators are independent tuning parameters. In the first step, the plugin optimizes the sieving buffer size for non-collective IO call sites and the collective buffer size for collective IO call sites. In the second tuning step, the plugin tunes the aggregator number for two-phase IO and collective call sites.

3.7 Compiler Flags

The *Compiler Flags Selection (CFS) plugin*, developed at Technical University of Munich (TUM), determines the best combination of compiler switches for a given application. The switches to be explored are given in a configuration file. The plugin applies expert knowledge in several ways to reduce the search space.

Since its evaluation is based on a recompilation of the application with the selected number of switches, it is important to reduce the compilation time. This is either done by the user giving the important files to recompile or the plugin determines the files automatically based on profiling results of the application.

Besides the exhaustive search strategy, the plugin provides the individual search strategy where the switches are evaluated in the order given by the configuration file. The breadth of the search can be controlled by the so called keep factor. It determines how many best settings of the last tuning parameter are taken into account in the evaluation of the next compiler switch.

Finally, the plugin provides a machine learning component, that is trained by collecting results from previous tunings of applications. The gathered information is used to compute probabilistic distributions for the values of the switches for the given application. Settings that were successful for similar applications have a higher probability than others. These distributions are then used in a probabilistic random search.

4   Tuning Plugin Interface (TPI)

All PTF's tuning plugins implement the *Tuning Plugin Interface (TPI)*. The frontend includes a driver that triggers these functions in a certain order.

The plugin interface is shown in Fig. 3. The frontend first calls `initialize` and provides a context object and the set of scenario pools used to manage scenarios from their creation until their evaluation was finished.

The driver context provides basic information, like the SIR file (introduced in Sect. 2), the number of threads and MPI processes specified at the command line, or general information (e.g. whether the program is instrumented or not). It also provides a timer which can be used to limit the search to a predefined search time.

The plugin can go through any number of tuning steps. The frontend triggers TPI functions in accordance to the general flow described in Fig. 4. Here is a short description of each of the interface methods in the base plugin class:

```
virtual void initialize(DriverContext *context,
                        ScenarioPoolSet *pool_set) = 0;
virtual void startTuningStep(void) = 0;
virtual bool analysisRequired(StrategyRequest** strategy) = 0;
virtual void createScenarios(void) = 0;
virtual void prepareScenarios(void) = 0;
virtual void defineExperiment(int numprocs,
                              bool *analysisRequired,
                              StrategyRequest** strategy) = 0;
virtual bool restartRequired(string *env, int *np,
                             string *cmd, bool *instrumented) = 0;
virtual bool searchFinished(void) = 0;
virtual void finishTuningStep(void) = 0;
virtual bool tuningFinished(void) = 0;
virtual Advice *getAdvice(void) = 0;
virtual void finalize(void) = 0;
virtual void terminate(void) = 0;
```

Fig. 3 Specification of the Tuning Plugin Interface functions

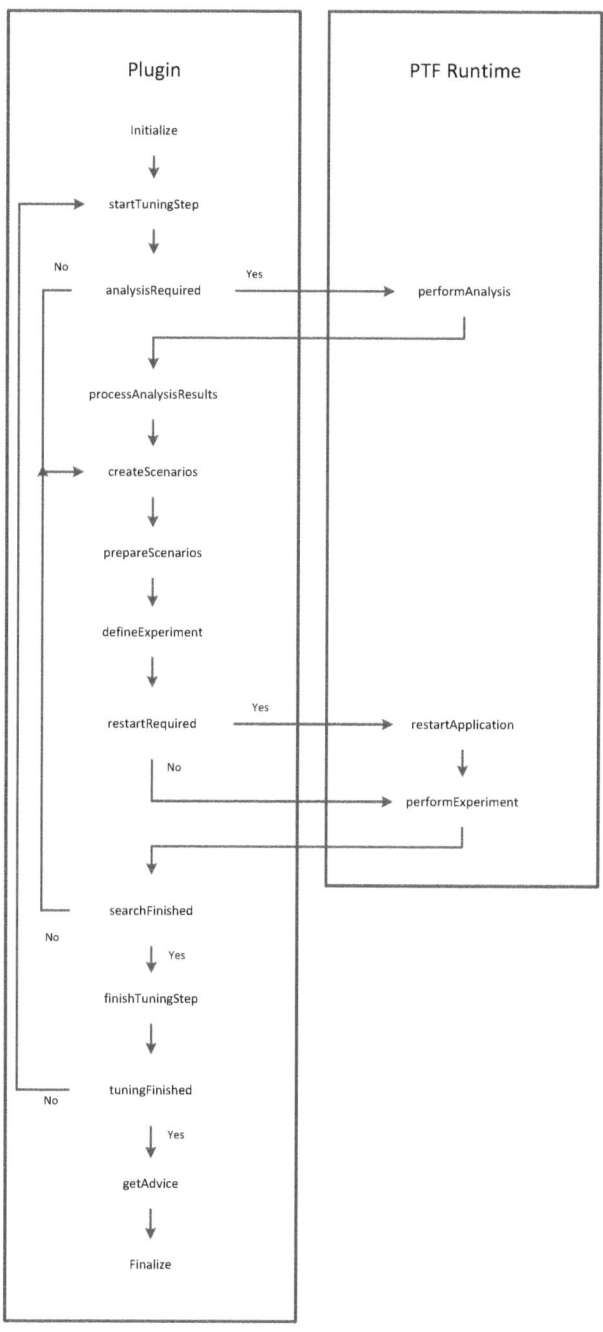

Fig. 4 Plugin flow chart and PTF runtime interaction

- **initialize**: All operations that are done once should be performed here. These include: memory allocations, required objects instantiation, storing the `context` and `pool_set` addresses passed to the plugin by the runtime system, etc.
- **startTuningStep**: Perform tuning step specific initialization. For example, set up relevant tuning parameters and variant spaces.
- **analysisRequired**: Indicates whether a pre-analysis is to be executed. Properties of the application can be determined before any tuning is performed. For example, code regions that take a significant amount of execution time and thus are good candidates for further tuning can be determined with a pre-analysis step.
- **createScenarios**: Creates a set of scenarios based on the pre-analysis results and a search algorithm.
- **prepareScenarios**: Executes pre-execution tuning actions. For example, the code can be recompiled if necessary.
- **defineExperiment**: Selects scenarios for the next experiment. It also can request the execution of a Periscope analysis strategy during the execution of the experiment. For example, a plugin can request an MPI analysis during the evaluation of a scenario to check which MPI calls have significant waiting time, if a certain tuning action is applied.
- **restartRequired**: Determines whether to restart the application possibly with other command line parameters. Different number of MPI processes can be requested, if necessary.
- **searchFinished**: Checks whether the search in this tuning step is finished or a new set of scenarios is to be created. For example, if a genetic search algorithm is used, it goes through a sequence of generations per search cycle.
- **finishTuningStep**: Finalizes the tuning step. Any required post-processing of the results of the particular tuning step takes place in this operation.
- **tuningFinished**: Checks, whether the plugin requires a next tuning step. As an example, the Master-Worker plugin (Sect. 3.4) splits the tuning into two tuning steps, the first for the partitioning factor and the second for the number of workers.
- **getAdvice**: Returns the tuning advice based on the findings of the plugin. Each plugin attempts to generate optimal settings for the tuning parameters of the application. The found combinations are presented to the user as a summary on the screen, and in complete form in a file in a specified XML format.
- **finalize**: Finish the plugin normally. Clean up memory and close file descriptors, etc.
- **terminate**: Terminate the plugin safely in case of error. This will be called in case of errors and does not appear in Fig. 4.

The frontend triggers `prepareScenarios`, `defineExperiment`, and `restartRequired` as often as necessary to execute all the created scenarios. Only after all created scenarios were evaluated, `searchFinished` is called to check whether this tuning step needs to create additional scenarios.

The scenarios are managed in a set of pools. First, all the created scenarios are inserted into the *Created Scenario Pool (CSP)*. During `prepareScenarios` some scenarios are popped, prepared, and inserted into the *Prepared Scenario Pool (PSP)*. In the `defineExperiment` call, the plugin decides which prepared scenarios can be run in a single experiment and are thus forwarded into the *Experiment Scenario Pool (ESP)*. Once they are executed, the scenarios are moved to the *Finished Scenario Pool (FSP)*. All these pools are part of a pool set that is passed to the plugin when it is initialized.

The execution of a plugin's search is controlled by the search strategy used. In addition, the plugin can define a timer giving a certain time budget for the search. This timer can be configured to call a function provided by the plugin when it expires. This function can then, for example, ensure that the plugin terminates without experimenting with any additional scenarios. The timer functionality is part of the plugin context.

A PTF plugin is a shared library that contains a specific implementation of the Tuning Plugin Interface (TPI), as well as management calls that provide information and allow it to be instantiated.

The TPI allows the plugin to define what is done in the operations of PTF's generic tuning flow. Some of the operations are optional, while many are mandatory. In the case of optional operations, empty implementations with default return values are sufficient. The flow chart of the generic tuning flow of PTF is presented in Fig. 4. The labeled enclosures are used to indicate which operations take place in the plugin and which take place in the Periscope runtime system. The operations that are in the center column are performed by PTF plugins, while the operations on the right are performed by the runtime system.

4.1 Plugin Management Operations

In addition to the Tuning Plugin Interface (TPI), management calls need to be included with every plugin implementation. These are C functions, since they have to be in the global scope without any C++ mangled names. Here is a short description about each C call:

- **getPluginInstance**: Returns an instance of the plugin.
- **getVersionMajor**: Returns the digit on the left of the dot of the plugin's version, as an integer type.
- **getVersionMinor**: Returns the digit on the right of the dot of the plugin's version, as an integer type.
- **getName**: Returns a string with the name of the plugin.
- **getShortSummary**: Returns a string with a short description of the plugin. This is recommended to be kept as a single line.

4.2 Data Model

In this section, a brief introduction to the core classes of the framework is presented. The classes in the data model of PTF are mostly data objects used by the plugins to tell the runtime system what to do.

4.2.1 Scenario Class

A PTF plugin needs to tell the framework what to set, where to set it, and what should be measured after that. This transfer of information is achieved through the Scenario data structure. It can be said that the main task of the PTF plugins is to create and track Scenario objects.

The Scenario class contains the following members:

- **id**: A unique identifier.
- **description**: A string with a short description.
- **tuned_region**: The code Region where it applies.
- **ts**: A *TuningSpecification* list.
- **pr**: A *PropertyRequest* list.
- **results**: The measurements from its evaluation.

The id is a unique number given to the Scenario once created and is used to track it. The description is an optional string that describes what the Scenario represents. The tuned_region is the location where the Scenario can be applied. The rest of the members require a more detailed description.

The *TuningSpecification* objects contain information about concrete values and location of parameters used for tuning. The *TuningSpecification* class contains the following members:

- **variant**: Set of concrete values to be written to the relevant *TuningParameter* objects.
- **context**: The context (location) where the variant will be applied.
- **ranks**: Ranks where the variant should be applied.

The variant is an instance of the Variant class. This class is used to represent concrete values for tuning parameters (represented by the *TuningParameter* class); a variant is in essence a point in a search space. The context is an instance of the *VariantContext* class. A *VariantContext* instance contains location information; locations can be files or code regions. Finally, the ranks member is an instance of the Ranks class. An instance of this class can represent all ranks, a set of ranks or a set of rank ranges.

The next member of the Scenario class in the list of *PropertyRequest* instances. As its name suggests, it contains a list of PTF properties to be collected by the experiments. This class contains the following members:

- **property_ids**: List of ID numbers of the properties to collect.

- **ranks**: Ranks where the properties should be collected. Object of the same type as in the *TuningSpecification*.
- **regions**: List of code locations where these properties are to be collected.

4.3 ScenarioPool Class

The *ScenarioPool* class is a collection of Scenario instances. These pools are used to coordinate work between the runtime system, the plugins and the search algorithms; these pools are core components of the tuning framework.

There are four such pools to be manipulated by the plugin (in the creation, preparation and experiment definition process) and the runtime system (in the evaluation process):

- **Created Scenario Pool (CSP)**: Contains all created scenarios in a search step.
- **Prepared Scenario Pool (PSP)**: Scenarios that are prepared are moved here from the CSP. Preparation can be application re-compilation, environment setup, etc.
- **Experiment Scenario Pool (ESP)**: Scenarios that were mapped to the execution environment (cores, threads, regions, etc.) and are ready for the next experiment are moved from the PSP to this pool.
- **Finished Scenario Pool (FSP)**: When the relevant experiment is finished and results collected, a Scenario is moved from the ESP to this pool.

4.4 Performance Properties

In the Periscope Tuning Framework (PTF), the result of experiments or analyses requested by the plugins are performance properties. A performance property characterizes a specific performance behavior, such as network performance, load imbalances, cache miss rates, etc. Performance properties can be specific to the type of parallel application, such as OpenMP or MPI specific properties. Properties can also be hardware specific, for example, they can belong to a specific Intel CPU or IBM Power CPU family. They can also be common across architectures, such as cache miss rates at different levels.

A property instance will contain the following members:

- **Condition**: Indicates whether the property was found in the program.
- **Confidence**: Degree of certainty on the condition as a percentage.
- **Severity**: Importance of the property, as a percentage.

The severity is the most relevant one when tuning automatically, since a higher or lower severity is a direct indication of the quality of the results gathered through an experiment.

4.5 Results Pools

Once the experiments required by specific scenarios and analyses requested by a plugin are completed, the results are collected and the scenarios are moved to the *Finished Scenario Pool (FSP)* from the *Prepared Scenario Pool (PSP)*. The collection of properties, that result from the Scenario instances' *PropertyRequest* lists, are placed into two result pools:

- **Analysis Results Pool (*ARP*)**: Contains properties that were collected as part of PTF analyses.
- **Scenario Results Pool (*SRP*)**: Properties that were collected as requested by Scenario instances are placed here.

These pools provide convenience methods for accessing the results (performance properties) based on common criteria, such as the Scenario ID or the search step where the relevant experiment or analysis took place.

5 Development of Tuning Plugins

This section describes how to initially configure the Periscope Tuning Framework (PTF) for plugin development, as well as how to utilize the development specific functionality to write an automatic tuning plugin.

5.1 Initial Configuration in Developer Mode

Extra features for plugin developers are included with the source code distribution of the framework. To enable these features on a particular build, the `--enable-developer-mode` option needs to be passed to the `configure` script. This flag will enable scripts that help in the generation of skeleton plugins and search algorithms, as well as provide the location of source code and installation locations for the different components of the framework.

The *configure* script distributed with the framework follows the conventions of the GNU coding standards [5]. In general, it is recommended to have a separate directory for the build, and that the *configure* script be called from it. The following is an example call to *configure* with the developer features enabled and with a custom install location:

```
<source-code-location>/configure
                --prefix=<install-location> --enable-developer-mode
```

In this case, PTF will be installed under the directory `install-location`. The distributed source code is located under the directory `source-code-location`; this can be specified as a relative or absolute path. If the `--prefix` option is not specified, the tool will be installed in the default location (refer to the GNU [5] conventions for a list of defaults). For more information about available options, pass the `--help` option to the *configure* script. After configuration, the framework can be built and installed by calling *make*.

The build system of the framework generates Makefiles that work well with the `-j` option of *make*. This flag specifies the number of threads used in the build process and can greatly accelerate the compilation. The following is an example call to *make* with n number of build threads:

```
make install -j <n>
```

The above command will build and install the framework in a single operation. After building and installing with development features enabled, developers are ready to start implementing automatic tuning plugins for the Periscope Tuning Framework (PTF).

5.2   Generating Empty Plugins with Stubs

As part of the Periscope Tuning Framework's source distribution, a set of skeletons are provided for developers to get started on automatic tuners. Currently, a skeleton for search algorithms and a skeleton for plugins are provided. Developers may want to start with a clean plugin that is essentially a clone of the provided skeleton, with modified names for the class and target library names. To achieve this, a convenience script has been distributed with the framework. If the PTF installation is up to date (and configured with the `--enable-developer-mode` option set), it can be verified that the generator script is available in the path:

```
which psc_generate_bare_plugin_from_skeleton
```

If the script is correctly installed, then you can proceed to generate a new plugin from the PTF skeleton. For example, to generate a minimal plugin where its desired class name is Foo and the target library is foo (to be passed to the `psc_frontend` at runtime, with the `--tune=foo` option), execute the following command:

```
psc_generate_bare_plugin_from_skeleton -c Foo -l foo
```

If there were no errors during this operation, the expected output is presented in Fig. 5. The `<path-to-source>` and `<version>` values should match the

```
Attempting to generate a new plugin with the provided information:
Class name: Foo
Library name: foo

PTF source code location:
<path-to-source>
PTF version:
<version>

Generating new plugin at <path-to-source>/autotune/plugins/foo...

Making the initial skeleton copy...

Processing the class header file...

Processing the class implementation file...

Processing the Makefile.am file...

Adding the new plugin to the build system...

Finished generating the new plugin.
Once installed (after 'make install' on your build directory) it should
be available with the 'psc_frontend' command by passing the option:
'--tune=foo' .
```

Fig. 5 Output while generating plugin skeleton

location and versions of the PTF source tree being used. A common error occurs when the desired directory already exists:

```
ERROR: <path-to-source>/autotune/plugins/foo exists.
                              Aborting the plugin creation process.
```

The issue can arise due to previous files from earlier runs of the script, or due to a name conflict with one of the available plugins of PTF. To check already used names (for the target library of the plugin) list the contents of the `autotune/plugins/` directory under the PTF's source directory.

The `psc_generate_bare_plugin_from_skeleton` script generates the class header, the C++ implementation file, its Makefile.am and the required directories. Additionally, the script adds the required include line into `autotune/plugins/Makefile.am`. For the removal of the plugin, the plugin directory with these files needs to be removed and the relevant line in `autotune/plugins/Makefile.am` deleted by the developer.

Once the skeleton has been generated, the build system should be able to compile and install the plugin in the configured install directory (with the `--prefix=<install-path>` option of the configure script). After a successful *make install* executed from the build directory, the foo directory should appear under `<install-path>/plugins/foo`.

The new plugin should now be loadable with the `--tune=foo` option of the `psc_frontend` program. The source code for the new generated plugin should be found under the directory `<path-to-source>/autotune/plugins/foo`.

```
#ifndef FOO_PLUGIN_H_
#define FOO_PLUGIN_H_

#include "AutotunePlugin.h"

class Foo: public IPlugin {

public:
  void initialize(DriverContext *context,
      ScenarioPoolSet *pool_set);
  void startTuningStep(void);
  bool analysisRequired(StrategyRequest** strategy);
  void createScenarios(void);
  void prepareScenarios(void);
  void defineExperiment(int numprocs, bool *analysisRequired,
      StrategyRequest** strategy);
  bool restartRequired(string *env, int *np,
      string *cmd, bool *instrumented);
  bool searchFinished(void);
  void finishTuningStep(void);
  bool tuningFinished(void);
  Advice *getAdvice(void);
  void finalize(void);
  void terminate(void);
};

#endif
```

Fig. 6 Generated header for plugin class Foo

The generated header file Foo.h under its `include/` folder should match the source code presented in Fig. 6.

In case of adding new headers and implementation files, the plugin's Makefile.am should be updated. This file is found under `src/`.

Additionally, the programmer should verify that the Foo.cc file under `src/` contains the required stubs for the Foo class specified in the above header.

5.3   OpenMP Scalability Example

Annotating code with OpenMP (http://openmp.org/) pragmas allows for the conversion of serial programs to parallel ones through multi-threading. After compiling a program with OpenMP enabled, the number of threads can be modified; the number of threads is a runtime option, not a compiled one. In this section, a plugin that performs basic scalability tests with OpenMP is presented. If the PTF installation is configured with the `--enable-developer-mode` then the plugin will be built and installed, as part of the tutorial plugins. Some sample output of the plugin, being

```
Periscope Performance Analysis Tool (ver. 1.1.0)
[psc_frontend][INFO:fe] Preparing to start the performance analysis...

Loaded Autotune components:

Plugin:          TutScalabilityBasic
Version:         1.0
Description:     Explores scalability of OpenMP codes with exhaustive search.

Search Algorithm: Exhaustive Search
Version:         1.0
Description:     Explores the full space spanned by all tuning parameters.

[psc_frontend][INFO:fe] Starting agents network...
[psc_frontend][INFO:fe] Starting application ./bt-mz.C.1 using 1 MPI procs
  and 4 OpenMP threads...


NAS Parallel Benchmarks (NPB3.3-MZ-MPI) - BT-MZ MPI+OpenMP Benchmark

Number of zones:  16 x  16
Iterations: 200    dt:   0.000100
Number of active processes:    1

Use the default load factors with threads
tal number of threads:     4  (  4.0 threads/process)

Calculated speedup =     4.00

Time step    1
[psc_frontend][INFO:fe] Prepared scenario pool not empty, still searching...
[psc_frontend][INFO:fe] Prepared scenario pool not empty, still searching...
[psc_frontend][INFO:fe] Prepared scenario pool not empty, still searching...
Optimum Scenario: 3


All Results:
Scenario        | Threads       | Time          | Speedup
0               |  1            |  3.42629      |  1
1               |  2            |  1.99972      |  1.71338
2               |  3            |  1.40804      |  2.43338
3               |  4            |  1.15698      |  2.96141


------------------------

[psc_frontend][INFO:fe] Plugin advice stored in: advice_17612.xml

----------------
Finished Periscope run! Search took 10.6839 seconds ( 0.0350299 seconds for startup  )
----------------
[psc_frontend][INFO:fe] Experiment completed!
[psc_frontend][INFO:fe] Exporting results to properties_tune_17612.psc
```

Fig. 7 Output of the sample plugin running one of the NAS Parallel Benchmarks

run with the multi-zone version of the NAS Parallel Benchmark's block-tridiagonal solver, is presented in Fig. 7.

The remainder of the section will be dedicated to describing this plugin's implementation. In each subsection, the source code of one interface method is presented and then explained. The operations are organized in the order that they are called from the runtime system.

```
void TutScalabilityBasic::initialize(DriverContext *context,
    ScenarioPoolSet *pool_set) {
  psc_dbgmsg(PSC_SELECTIVE_DEBUG_LEVEL(AutotunePlugins),
                "TutScalabilityBasic: call to initialize()\n");

  this->context = context;
  this->pool_set = pool_set;

  TuningParameter *numberOfThreadsTP = new TuningParameter();
  numberOfThreadsTP->setId(0);
  numberOfThreadsTP->setName("NUMTHREADS");
  numberOfThreadsTP->setPluginType(TUTSCALABILITYBASIC);
  numberOfThreadsTP->setRange(1, context->getOmpnumthreads(), 1);
  numberOfThreadsTP->setRuntimeActionType(FUNCTION_POINTER);
  tuningParameters.push_back(numberOfThreadsTP);

  int major, minor;
  string name, description;
  context->loadSearchAlgorithm("exhaustive",
                        &major, &minor, &name, &description);
  searchAlgorithm = context->getSearchAlgorithmInstance("exhaustive");
  if (searchAlgorithm){
    print_loaded_search(major, minor, name, description);
    searchAlgorithm->initialize(context, pool_set);
  }
}
```

Fig. 8 Plugin initialization with a loaded search algorithm

5.3.1 Plugin Initialization

The source code for this operation is presented in Fig. 8. As can be seen, the very first operation is a debugger output call. To allow PTF to selectively display debug messages that are produced by plugins, the developer should set the PSC_SELECTIVE_DEBUG_LEVEL(AutotunePlugins) debug level as the first parameter of the psc_dbgmsg call. This debug level can then be selected at runtime by passing --selective-debug=AutotunePlugins to the *psc_frontend* command.

The next two operations set the local context and pool_set references with the ones passed by the runtime system. This must be done by all plugins. The context object is an instance of the *DriverContext* class, that is instantiated by the runtime system and is used to offer services to the plugins. The pool_set contains the scenario pools and results pools described in Sect. 4.3.

The next step is to define a tuning parameter (instance of the *TuningParameter* class). These parameters represent aspects of the application that can affect its performance and can be modified. In the case of this plugin, the parameter represents the number of OpenMP threads that will be created at runtime. After creating the object, its ID number, name, type, range and action type are set. The number of threads specified through the command line options is acquired, and then set as the maximum in the tuning parameter's range. This is indeed a sensible way of setting the maximum, but there are other good alternatives that developers can use, such as the maximum number of cores or hardware threads in the system. Finally, the new parameter is pushed to the tuningParameters vector. This C++ std::vector was defined in the plugin's header (under `autotune/plugins/tutorials/ include/TutScalabilityBasic.h`).

The final task done at `initialize` is the creation of a search algorithm instance. There are multiple search algorithms available in the PTF distribution. Developers are also free to develop their own. Once a search algorithm has been made available in the framework, it can be loaded through one of the services provided through the `context`. This is done in two steps: first, the implementation is loaded with the `loadSearchAlgorithm` call, and second, an instance is acquired with the `getSearchAlgorithmInstance` call. Once the search algorithm instance is loaded, its information is printed and it is initialized with the same `context` and `pool_set` that was provided to the plugin by the runtime system.

5.3.2 Start Tuning Step

The source code for this operation is presented in Fig. 9. This plugin contains a single tuning step. Before this step can be performed, the variant space and the search space need to be created and configured for the search algorithm. First, the VariantSpace and SearchSpace instances are created. After that, the `tuningParameters` vector is iterated and each parameter is added to the variant space. Only one parameter is added to this vector, but iterating is good practice since it allows for code reuse if the plugin is extended. Once the `variantSpace` is ready, it is set in the `searchSpace` instance. The SearchSpace class consists of its `VariantSpace` and additional information, such as restrictions and the location where the variants should be applied. In this case, the location where to apply it is set as the phase region. The phase region is either the `main` procedure of a program, or a user region defined with PTF pragmas. Once the search space is set up, it is passed to the search algorithm with the *addSearchSpace* call.

```
void TutScalabilityBasic::startTuningStep(void) {
  VariantSpace *variantSpace=new VariantSpace();
  SearchSpace *searchSpace=new SearchSpace();

  for (int i = 0; i < tuningParameters.size(); i++) {
    variantSpace->addTuningParameter(tuningParameters[i]);
  }
  searchSpace->setVariantSpace(variantSpace);
  searchSpace->addRegion(appl->get_phase_region());

  searchAlgorithm->addSearchSpace(searchSpace);
}
```

Fig. 9 Example implementation of *startTuningStep()*

```
bool TutScalabilityBasic::analysisRequired(StrategyRequest** strategy) {
  return false;
}
```

Fig. 10 Minimal implementation of *analysisRequired()*

```
void TutScalabilityBasic::createScenarios(void) {
  searchAlgorithm->createScenarios();
}
```

Fig. 11 Example of *createScenarios()* with a loaded search algorithm

```
void TutScalabilityBasic::prepareScenarios(void) {
  while(!pool_set->csp->empty()) {
    pool_set->psp->push(pool_set->csp->pop());
  }
}
```

Fig. 12 Minimal implementation of *prepareScenarios()* (no preparation)

5.3.3 Analysis Required

The relevant source snippet for this operation is presented in Fig. 10. In this particular plugin, no analysis is performed on the application before the search. Therefore, returning false is all that has to be done here.

5.3.4 Create Scenarios

The relevant scenario creation code is shown in Fig. 11. Because a search algorithm (with an instance reference in the *searchAlgorithm* pointer) was loaded, the *createScenarios()* operation should be offloaded to it. The search algorithm will generate the scenarios and put them in the *Created Scenario Pool (CSP)*.

5.3.5 Prepare Scenarios

The scenarios created require no preparation; they should simply be moved to the *Prepared Scenario Pool (PSP)*. The minimal required code for this operation is presented in Fig. 12. If the thread number would have to be changed through the OpenMP environment variable OMP_NUM_THREADS, then it would have to be updated here. The framework has support for thread changes in the runtime system, so no environment preparation needs to take place here.

```
void TutScalabilityBasic::defineExperiment(int numprocs,
    bool *analysisRequired, StrategyRequest** strategy) {
  Scenario *scenario = pool_set->psp->pop();
  scenario->setSingleTunedRegionWithPropertyRank(
    appl->get_phase_region(), EXECTIME, 0);
  pool_set->esp->push(scenario);
}
```

Fig. 13 Example of *defineExperiment()* with a execution time objective

```
bool TutScalabilityBasic::restartRequired(string *env,
    int *numprocs, string *command, bool *is_instrumented) {
  return false;
}
```

Fig. 14 Minimal implementation of *restartRequired()*

5.3.6 Define Experiment

The experiment definition code is presented in Fig. 13. When an experiment is defined, the Scenarios are mapped to the runtime environment (processes, threads and code locations). In this example, one scenario at a time is taken, and its objective is set as the execution time property of the phase region. This is requested with the *setSingleTunedRegionWithPropertyRank()* call of the Scenario class.

Once the scenario is set up, it is pushed to the *Experiment Scenario Pool (ESP)*, and the runtime system will perform the required experiments to acquire its properties (only the execution time in this case).

5.3.7 Restart Required

For the evaluation of the Scenario that was pushed to the *ESP*, no restart is required because the number of OpenMP threads can be changed at runtime; therefore, returning false here is sufficient. The minimal code required for this operation is presented in Fig. 14.

5.3.8 Search Finished

Since the plugin relies on a search algorithm for Scenario creation, it also delegates this decision to it. The search algorithm will take a look at the results pool and make a decision. In the case of the ExhaustiveSearch algorithm loaded in this plugin, this call always returns true since all possible Scenarios are created at the first call of its `createScenarios` method. The implementation of this operation is presented in Fig. 15.

```
bool TutScalabilityBasic::searchFinished(void) {
  return searchAlgorithm->searchFinished();
}
```

Fig. 15 Example of *searchFinished()* with a loaded search algorithm

```
void TutScalabilityBasic::finishTuningStep(void) { }
```

Fig. 16 Minimal implementation of *finishTuningStep()*

```
bool TutScalabilityBasic::tuningFinished(void) {
  return true;
}
```

Fig. 17 Minimal implementation of *tuningFinished()*

5.3.9   Finish Tuning Step

There are no operations to be done at the end of this plugin's single tuning step. An empty implementation here is sufficient and is shown in Fig. 16.

5.3.10   Tuning Finished

In this plugin, the search is performed in a single tuning step. Returning `true` here is sufficient. The implementation is presented in Fig. 17.

5.3.11   Get Advice

The code to generate the advice to the user is presented in Fig. 18. This operation presents information to the user and generates XML output for further processing.

After a few declarations, the ID of the scenario with the best time is collected from the search algorithm and presented. This is done with the *getOptimum()* call. After that, all results are presented in tabular form. All scenarios in the *Finished Scenario Pool (FSP)* are iterated and for each the ID, number of threads, execution time and speed up over the single threaded case is presented.

Finally, an Advice instance is built that is passed to the runtime system. This object is then used by the runtime system to generate the XML output of the advice.

5.3.12   Finalize

All created objects need to be freed in this operation. Loaded components through the `context` will be unloaded by the runtime system, but obtained instances must be deleted here. The `pool_set` is also handled by the runtime system. The relevant source code is presented in Fig. 19.

```
Advice *TutScalabilityBasic::getAdvice(void) {
  std::ostringstream result_oss;
  map<int,double> timeForScenario=searchAlgorithm->getSearchPath();
  double serialTime = timeForScenario[0];

  int optimum = searchAlgorithm->getOptimum();
  result_oss << "Optimum Scenario: " << optimum << endl << endl;

  result_oss << "\nAll Results:\n";
  result_oss << "Scenario\t|  Threads\t|  Time\t|  Speedup\t\n";

  for (int scenario_id = 0;
       scenario_id < pool_set->fsp->size(); scenario_id++) {
    Scenario *sc = (*pool_set->fsp->getScenarios())[scenario_id];
    list<TuningSpecification*>* tuningSpec =
                               sc->getTuningSpecifications();
    map<TuningParameter*,int> tpValues =
      tuningSpec->front()->getVariant()->getValue();
    int threads = tpValues[tuningParameters[0]];
    double time = timeForScenario[scenario_id];

    result_oss << scenario_id << "\t\t|   " << threads << "\t\t|   "
      << time << "\t|   " << serialTime/time << endl;
  }
  result_oss << "\n------------------------" << endl << endl;

  cout << result_oss.str();

  map<int, Scenario*>::iterator scenario_iter;
  for(scenario_iter = pool_set->fsp->getScenarios()->begin();
      scenario_iter != pool_set->fsp->getScenarios()->end();
        scenario_iter++){
    Scenario *sc=scenario_iter->second;
    sc->addResult("Time", timeForScenario[sc->getID()]);
  }

  Scenario *bestScenario= (*pool_set->fsp->getScenarios())[optimum];
  return new Advice(getName(), bestScenario, timeForScenario,
    "Time", pool_set->fsp->getScenarios());
}
```

Fig. 18 Example implementation of *getAdvice()* using an *Advice* constructor

```
void TutScalabilityBasic::finalize() {
  delete searchAlgorithm;
  for (int i = 0; i < tuningParameters.size(); i++) {
    delete tuningParameters[i];
  }
}
```

Fig. 19 Example of *finalize()*. Allocated memory must be deleted in this operation

5.4   *Vector Range Restriction*

In this section, the reader's understanding on the concept of a *TuningParameter* is extended, and then a description of how to restrict the possible values that a *TuningParameter* instance can take is given. Finally, the plugin (started in Sect. 5.3) will be extended in its *initialization()* step, to make use of the *Restriction* class of the framework.

5.4.1 Tuning Parameter

The *TuningParameter* class represents a range of integer values that affect performance. In this case, these parameter represents the number of OpenMP threads in a parallel region. The *TuningParameter* class has the following members:

- **ID**: A unique identifier.
- **runtimeActionType**: The runtime tuning action type. A tuning action is the operation that is required to set the concrete value given to the parameters in a variant (a value between the *from* and *to* value that fits the *step* size and/or the *Restriction*).
- **name**: A name given to it.
- **from**: The lowest possible value of the parameter.
- **to**: The highest possible value of the parameter.
- **step**: The value that is used to increment from the *from* value until the *to* value.
- **restrictions**: This further restricts the possible values that the parameter can take. This is described in the next subsection.

The *TuningParameter* instances are aggregated as concrete values into a vector class called Variant. A Variant therefore represents a point in the search space; a unique combination of parameters that may result in better or worse performance. In that sense, a *TuningParameter* instance represents a dimension in a search space.

5.4.2 Initialization

The *Restriction* member will be further explained here. This member can be accessed through getters and setters and allows to restrict the possible values that the parameter can take. If the *from*, *to*, and *step* are sufficient for a particular plugin's use case, then the *Restriction* member can be left unset. This section shows how to set it to powers of 2 between the *TuningParameter* value boundaries. The relevant code snippet is presented in Fig. 20. The relevant changes are indicated by the comment lines.

First, a new *Restriction* object is created. After that, a counter is incremented in powers of 2 and its value is added as an element of the *Restriction* instance. The region and the type are then set to some default value, and finally, this restriction is set on the *TuningParameter* instance. This restricts the *TuningParameter* so that it only takes power of 2 values for the OpenMP threads value.

5.5 *Working with Regions*

In PTF, a region represents a location in the source code. Regions are static information collected by the instrumenter and stored in the SIR file (in XML format). Region information can be obtained by the plugins and used to target

```
void TutVectorRange::initialize(DriverContext *context,
    ScenarioPoolSet *pool_set) {
  this->context = context;
  this->pool_set = pool_set;

  TuningParameter *numberOfThreadsTP = new TuningParameter();
  numberOfThreadsTP->setId(0);
  numberOfThreadsTP->setName("NUMTHREADS");
  numberOfThreadsTP->setPluginType(TUTVECTORRANGE);
  numberOfThreadsTP->setRange(1, context->getOmpnumthreads(), 1);

  // start of changes for the restriction
  Restriction *r = new Restriction();
  for (int i=1; i<=context->getOmpnumthreads(); i=i*2){
    r->addElement(i);
  }
  r->setRegion(NULL);
  r->setType(2);
  numberOfThreadsTP->setRestriction(r);
  // end of changes for the restriction

  numberOfThreadsTP->setRuntimeActionType(FUNCTION_POINTER);
  tuningParameters.push_back(numberOfThreadsTP);

  int major, minor;
  string name, description;
  context->loadSearchAlgorithm("exhaustive",
                                &major, &minor, &name, &description);
  searchAlgorithm = context->getSearchAlgorithmInstance("exhaustive");
  if (searchAlgorithm){
    print_loaded_search(major, minor, name, description);
    searchAlgorithm->initialize(context, pool_set);
  }
}
```

Fig. 20 Extended *initialize()* method with vector range restriction

optimizations only to relevant locations, depending on the aspect to be tuned in
the application.

All regions are stored in a tree structure (similar to an Abstract Syntax Tree, as
produced by compiler frontends). The phase region can be the *main()* procedure
of the application, if only automatic instrumentation has been performed. If the
user added a user region through the framework's pragmas, then this region will
be the *phase* region. This has consequences for the plugins, since static information
will be contained in the SIR file about all regions, but experiments will be started
either at the *main()* procedure or at the user region, depending on the situation.
Well developed plugins should take both of these situations into consideration and
adjust their mode of operation accordingly. In this section, the plugin is extended
to illustrate how to process region information. The changes are described in the
following subsections.

```
void TutRegions::initialize(DriverContext *context,
        ScenarioPoolSet *pool_set) {
  this->context = context;
  this->pool_set = pool_set;

  // added validation before processing region information
  if (appl->get_regions().empty()) {
    psc_errmsg(PSC_SELECTIVE_DEBUG_LEVEL(AutotunePlugins),
                "TutRegions plugin : No Region found. Exiting.\n");
    throw 0;
  }

  ...
}
```

Fig. 21 Extended *initialize()* method with region information

5.5.1 Initialization

In the initialization, the application object *app* is initialized. After that, a check is added to verify that the list of regions is not empty. This is in general good practice and a check prevents plugins that depend on region information from executing and terminating with an error after the application was launched. The relevant changes are presented in Fig. 21.

5.5.2 Start Tuning Step

To set up the single tuning step, the list of regions is obtained from the *app* object. Although in general not done in real plugins, an application specific selection of the region is shown here. The OpenMP parallel region that belongs to the file z_solve.f, of the NPB's BT-MZ benchmark, is detected. After that, a new search space and variant space are created. The *TuningParameter* instances are added to the variant space by iterating through them. Afterwards, the region information is printed to the screen and the region is added to the new search space instance. The parameters that are part of the search space are then set for this region, during the experiments. Finally, the search space is passed to the search algorithm, so that scenario creation and results processing operations are delegated to it. The relevant code is presented in Fig. 22.

5.5.3 Define Experiment

The last extension done to the plugin is a change in the experiment definition. The changes are presented in Fig. 23. Similarly to the previous step, the list of regions are acquired and the one that belongs to the file z_solve.f is selected. One scenario is then taken from the *Prepared Scenario Pool (PSP)* and its objective property is requested for this particular region only.

```
void TutRegions::startTuningStep(void) {
  // Tuning Parallel Regions
  std::list<Region*> code_regions;
  code_regions = appl->get_regions();
  std::list<Region*>::iterator region;

  // iterating over all regions
  for (region = code_regions.begin();
                  region != code_regions.end(); region++) {
    if ((*region)->get_type() == PARALLEL_REGION
                    || (*region)->get_type() == DO_REGION) {
      if((*region)->getFileName() == "z_solve.f" ) {
        SearchSpace *searchspace = new SearchSpace();
        VariantSpace *variantSpace = new VariantSpace();

        for (int i = 0; i < tuningParameters.size(); i++) {
          variantSpace->addTuningParameter(tuningParameters[i]);
        }
        searchspace->setVariantSpace(variantSpace);

        searchspace->addRegion(*region);
        searchAlgorithm->addSearchSpace(searchspace);
      }
    }
  }
}
```

Fig. 22 Extended *startTuningStep()* method with region selection

```
void TutRegions::defineExperiment(int numprocs,
          bool *analysisRequired, StrategyRequest** strategy) {
  Scenario *scenario = pool_set->psp->pop();

  // Tuning Parallel Regions
  std::list<Region*> code_regions;
  code_regions = appl->get_regions();
  std::list<Region*>::iterator region;

  // iterating over all regions
  for (region = code_regions.begin();
                  region != code_regions.end(); region++) {
    if ((*region)->get_type() == PARALLEL_REGION
                    || (*region)->get_type() == DO_REGION) {
      if((*region)->getFileName() == "z_solve.f" ) {
        scenario->setSingleTunedRegionWithPropertyRank(
                                (*region), EXECTIME, 0);
      }
    }
  }
  pool_set->esp->push(scenario);
}
```

Fig. 23 Extended *defineExperiment()* method with region specific objective

5.6   Mapping Scenarios to the Execution Environment

In this section, the *defineExperiment()* method of the plugin will be modified, so
that the available parallelism in the running job can be used to accelerate the search
process. In the plugin, a search is performed to find the best number of OpenMP
threads; the threads have been manipulated across the complete program, in a user

defined region, as well as in a specific parallel region (in the file z_solve.f). Now
another dimension of control will be added, by specifying specific thread counts per
rank.

5.6.1 Define Experiment

As can be seen in the code snippet in Fig. 24, the only difference in the *defineExperiment()* method is that the *TuningSpecification* instance of the scenario is updated.
The *TuningSpecification* is changed from the default (to apply the tuning action to
all ranks) to now apply the tuning action only to a specific rank. The property request
is also set to the same rank, through the *setSingleTunedRegionWithPropertyRank()*
call.

This is done for each rank in the process list; this allows the plugin to evaluate as
many scenarios as there are ranks, in parallel, in a single experiment.

```cpp
void TutMultiRank::defineExperiment(int numprocs,
                    bool *analysisRequired, StrategyRequest** strategy) {
  for(int i=0; i<numprocs; i++) {
    if(pool_set->psp->empty()) {
      break;
    }

    Scenario *scenario = pool_set->psp->pop();
    const list<TuningSpecification*> *ts =
                            scenario->getTuningSpecifications();
    ts->front()->setSingleRank(i);

    // Tuning Parallel Regions
    std::list<Region*> code_regions;
    code_regions = appl->get_regions();
    std::list<Region*>::iterator region;

    // iterating over all regions
    for (region = code_regions.begin();
                    region != code_regions.end(); region++) {
      if ((*region)->get_type() == PARALLEL_REGION
                    || (*region)->get_type() == DO_REGION) {

        if((*region)->getFileName() == "z_solve.f" ) {
          scenario->setSingleTunedRegionWithPropertyRank(
                                    (*region), EXECTIME, i);
        }
      }
    }
    pool_set->esp->push(scenario);
  }
}
```

Fig. 24 Extended *defineExperiment()* method with different tuning action per MPI rank

```
void TutMultipleRegions::startTuningStep(void) {
  std::list<Region*> code_regions;
  code_regions = appl->get_regions();
  std::list<Region*>::iterator region;

  // iterating over all regions
  int count = 0, parallel_regions = 0;
  for (region = code_regions.begin();
                       region != code_regions.end(); region++) {
    if ((*region)->get_type() == PARALLEL_REGION) {
      parallel_regions++;
      code_region_candidates[(*region)->getIdForPropertyMatching()]
                                                         = *region;
    }
  }
}
```

Fig. 25 Extended *startTuningStep()* method with parallel regions selection

5.7   Analysis and Region Selection

This section presents the usage of a PTF analysis in a plugin. It is shown how it can be used to detect areas of interest. The `Region` class is used to represent locations of interest in the source code of the application that is being tuned. In this section, it will be shown how to acquire region information and process it. Finally, an explanation on how to filter the *Region* instances to include only the ones that impact performance significantly is presented.

5.7.1   Start Tuning Step

In this example, the single tuning step approach is continued. The code snippet for its *startTuningStep()* method is presented in Fig. 25. In it, the list of regions extracted by the instrumenter is acquired from the SIR file, with the *get_regions()* method. Afterwards, only the OpenMP parallel regions are added to a list of candidates for tuning (this list of candidates is declared in the header of the plugin). The type of a *Region* instance can be inspected with the *get_type()* method.

5.7.2   Analysis Required

In order to obtain performance information about the OpenMP parallel regions, an analysis is requested. To perform an analysis, the plugin needs to return *true* and pass a *StrategyRequest* object by reference to the runtime system. This is done in the *analysisRequired()* method and the relevant code is presented in Fig. 26. To create this object, a new instance of the class *StrategyRequestGeneralInfo* is created and

```
bool TutMultipleRegions::analysisRequired(StrategyRequest** strategy) {
  StrategyRequestGeneralInfo* analysisStrategyGeneralInfo =
                                      new StrategyRequestGeneralInfo;
  std::map<string, Region*>::iterator region;

  analysisStrategyGeneralInfo->strategy_name = "ConfigAnalysis";
  analysisStrategyGeneralInfo->pedantic = 0;
  analysisStrategyGeneralInfo->delay_phases = 0;
  analysisStrategyGeneralInfo->delay_seconds = 0;

  PropertyRequest *req = new PropertyRequest();

  req->addPropertyID(EXECTIME);
  for (region = code_region_candidates.begin();
                region != code_region_candidates.end(); region++) {
    req->addRegion(region->second);
  }
  req->addAllProcesses();

  list<PropertyRequest*>* reqList = new list<PropertyRequest*>;
  reqList->push_back(req);

  *strategy = new StrategyRequest(reqList, analysisStrategyGeneralInfo);
  return true;
}
```

Fig. 26 Initial *initialize()* method with analysis strategy request

its members are filled in: the name, the pedantic flag, the number of delay phases and seconds. Here is a brief explanation of their meaning:

- **strategy_name**: Name of the strategy to perform.
- **pedantic**: If set, all properties are reported, regardless of their severity.
- **delay_phases**: Number of times to run the phase region before collecting data.
- **delay_seconds**: Number of seconds to run before collecting data.

After the *StrategyRequestGeneralInfo* object was ready, a new *PropertyRequest* instance is created. In this case, the plugin requests an *EXECTIME* property for the regions. After that, it adds all the candidate regions (the parallel ones that were collected previously) to the request. To request these properties from all processes, a call to its *addAllProcesses()* method is done. The *StrategyRequest* constructor requires a list of *PropertyRequest* objects and one StrategyRequestGeneralInfo object; therefore, a list is created, the single request is added to it and then the constructor is called with the StrategyRequestGeneralInfo instance.

After returning *true*, the runtime system proceeds to perform the analysis and will populate the *Analysis Results Pool (ARP)* with the resulting properties.

5.7.3 Create Scenarios

The extended implementation of *createScenarios()* is presented in Fig. 27. In this operation, the *SearchSpace* instance to pass to the search algorithm is set up, so that the plugin can delegate the creation and evaluation of scenarios to it.

```
void TutMultipleRegions::createScenarios(void) {
  list<MetaProperty>::iterator property;
  MetaProperty longest_running_property;
  list<MetaProperty> found_properties =
                  pool_set->arp->getPreAnalysisProperties(0);
  double severity = 0;

  SearchSpace *searchspace = new SearchSpace();
  VariantSpace *variantSpace = new VariantSpace();

  for (int i = 0; i < tuningParameters.size(); i++) {
    variantSpace->addTuningParameter(tuningParameters[i]);
  }
  searchspace->setVariantSpace(variantSpace);

  for (property = found_properties.begin();
                  property != found_properties.end(); property++){
    if (property->getSeverity() > severity){
      severity = property->getSeverity();
      longest_running_property = *property;
    }
  }
  selected_region =
        code_region_candidates[longest_running_property.getRegionId()];
  searchspace->addRegion(selected_region);

  searchAlgorithm->addSearchSpace(searchspace);
  searchAlgorithm->createScenarios();
}
```

Fig. 27 Extended *createScenarios()* method with region selection based on runtime

First, the *SearchSpace* and *VariantSpace* instances are created. The *TuningParameter* instances are added to the *VariantSpace* instance iteratively.

The properties are stored in the *AnalysisResultsPool* instance based on the tuning step number. In this example, only one tuning step (number zero) is done. To acquire these properties in a list, the *getPreAnalysisProperties()* method of the pool is called. After that, the list of properties that were collected by the runtime system as part of the previously requested analysis is iterated. The longest running property is collected during the iteration and its corresponding region is selected for tuning.

The selected *Region* is then set on the search space for the search algorithm. After that is done, the plugin can call the search algorithm's *createScenarios()* method, to delegate the creation of new scenarios. All new scenarios will be placed in the *Created Scenario Pool (CSP)* by the search algorithm.

5.7.4 Define Experiment

The *defineExperiment()* operation is also extended; the changes are presented in Fig. 28. To define the experiment, a single scenario is taken and its objective property set to *EXECTIME* in the previously selected region at rank 0. It is then moved to the *Experiment Scenario Pool (ESP)*. The runtime system takes the scenarios in that pool and evaluates them. These operations are repeated until the scenarios in the pools are all evaluated.

```
void TutMultipleRegions::defineExperiment(int numprocs,
                  bool *analysisRequired, StrategyRequest** strategy) {
  Scenario *scenario = pool_set->psp->pop();
  scenario->setSingleTunedRegionWithPropertyRank(
                                    selected_region, EXECTIME, 0);
  pool_set->esp->push(scenario);
}
```

Fig. 28 Extended *defineExperiment()* method with specific region objective

5.8   Experiments with Scenario Analysis

As has been covered earlier, the automatic tuning process with PTF relies on
the setting of *tuning parameters* in particular code *regions* and measuring the
effects through the collection of performance *properties*. These structures, used
to set parameters and measure effects, are contained in *Scenario* instances that
are evaluated in experiments. Along with the experiments specified through the
Scenarios, a plugin can request analysis strategies. Analysis strategies were covered
already in Sect. 5.7, where a pre-analysis is requested and its results used to select
the longest running *parallel region* of type OpenMP, and therefore the most relevant
for the automatic tuner. This section shows how to request for an analysis along
with an experiment, in order to collect additional performance properties under the
conditions set by a specific scenario.

5.8.1   Defining an Experiment with an Analysis

The only changes required take place at the *defineExperiment()* method; the relevant
code snippet is presented in Fig. 29. The operation starts with a *pop*, *configure* and
push sequence, just like in previous examples. In this case, the plugin selects the
phase region and requests its time of execution at rank zero. In addition to this, it
then needs to define the analysis. For this example, the regions of interest that were
detected in the pre-analysis are selected. Finally, the *StrategyRequest* instance is set
to this new analysis and the *analysisRequired* flag is set to true.

5.8.2   Providing Extra Information to the User

The *getAdvice()* method is updated to show the extra information that was collected
with the analysis, along with the experiment. For that, the best times per sub-region
are collected and presented to the user, as shown in the code snippet presented in
Fig. 30.

```
void TutScenAnalysis::defineExperiment(int numprocs, bool *analysisRequired,
                                       StrategyRequest** strategy) {
    Scenario *scenario = pool_set->psp->pop();
    scenario->setSingleTunedRegionWithPropertyRank(appl->get_phase_region(),
                                                   EXECTIME, 0);
    pool_set->esp->push(scenario);

    // requesting region specific properties
    StrategyRequestGeneralInfo* analysisStrategyRequest =
                                       new StrategyRequestGeneralInfo;
    analysisStrategyRequest->strategy_name = "ConfigAnalysis";
    analysisStrategyRequest->pedantic = 1;
    analysisStrategyRequest->delay_phases = 0;
    analysisStrategyRequest->delay_seconds = 0;

    list<PropertyRequest*>* reqList = new list<PropertyRequest*>;
    for(int i=0; i<noSigRegions; i++) {
        PropertyRequest *req = new PropertyRequest();
        req->addPropertyID(EXECTIME);
        req->addRegion(code_region_candidates[regionExectimeMap[i].first]);
        req->addSingleProcess(0);
        reqList->push_back(req);
    }

    StrategyRequest *sub_strategy = new StrategyRequest(reqList,
                                                  analysisStrategyRequest);
    (*strategy) = sub_strategy;
    (*strategy)->printStrategyRequest();
    (*analysisRequired) = true;
}
```

Fig. 29 Extended *defineExperiment()* method with an analysis request

```
    ...
    result_oss << endl << "Best configurations for individual regions:" << endl;

    for(int i=0; i<noSigRegions; i++) {
        result_oss << "Region: " << regionName[i] << " " ;
        result_oss << "Threads: " << regionBestThreads[i] << " ";
        result_oss << "ExecTime: " << regionExecTime[i] << endl;

        // accumulating best scenario description
        Scenario *sc = (*pool_set->fsp->getScenarios())[regionBestScenario[i]];
        string scenDescription = sc->getDescription();
        std::stringstream strStream;
        strStream << "Best scenario for region: " << regionName[i];
        if(scenDescription.empty()) {
            sc->setDescription(strStream.str());

        } else {
            scenDescription = scenDescription + " | " + strStream.str();
            sc->setDescription(scenDescription);
        }
    }
    ...
```

Fig. 30 Added detailed performance information output to *getAdvice()*

6 OpenMP Scalability Plugin Results

As part of the source code distribution of the Periscope Tuning Framework (PTF), a set of plugins and their documentation, as well as a complete tutorial covering the topics in this presentation and more (with partial implementations), are included. One of the distributed plugins implements many of the techniques presented in preceding sections: the OpenMP Scalability Plugin. In this section, single node results from this plugin, applied to several of the NPB benchmarks and run on the SuperMUC [1] system, are presented.

6.1 SuperMUC Nodes

The SuperMUC system contains two types of nodes referred to as thin and fat. The fat nodes contain 40 cores in total, from 10 core CPUs in a 4 socket configuration. The thin nodes contain 16 cores total, from 8 core CPUs in a 2 socket configuration. Both types of nodes have Intel's HyperThreading enabled. In depth detail about the system and these nodes can be found under http://www.lrz.de/services/compute/supermuc/systemdescription/.

6.2 Thin Node Scaling

The results of OpenMP scaling on the SuperMUC's thin nodes are presented in Fig. 31. As can be seen, there is acceptable scaling up to 16 threads; after that, the change to 32 threads still produce significant increases in performance but not close to doubling what is observed under the lower thread count changes. This is expected since in the last step 16 cores with 32 hardware threads are at work. With Intel's HyperThreading technology, the computational units and the caches are shared among hardware threads, so a doubling of performance is not realistic. From the figure, it can also be observed that scaling differs per application.

6.3 Fat Node Scaling

The results of OpenMP scaling on the SuperMUC's fat nodes is presented in Figs. 32 and 33. The plots with powers of 2 was included to allow direct comparisons to the results collected on the thin nodes. The scalability on these nodes is very different, when compared to the thin nodes, particularly on HyperThreading situations.

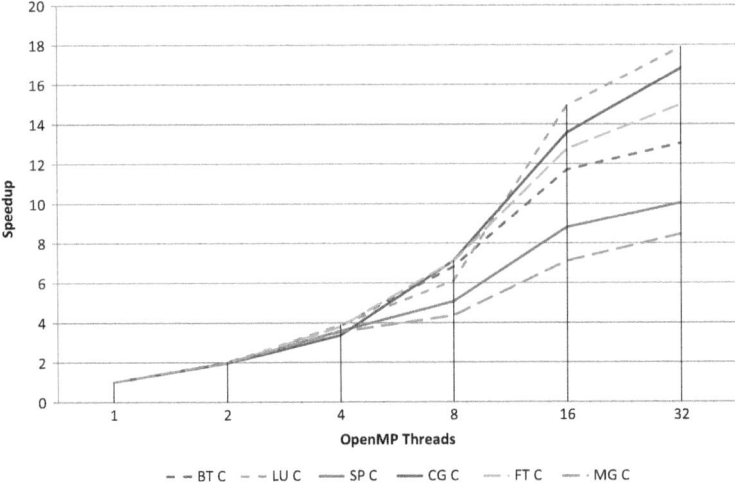

Fig. 31 SuperMUC thin node scaling, NPB MZ benchmarks (powers of 2 OpenMP threads)

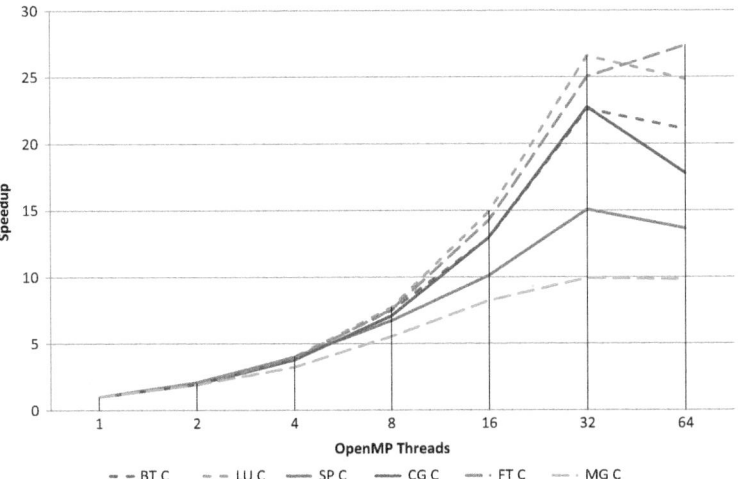

Fig. 32 SuperMUC fat node scaling, NPB MZ benchmarks (powers of 2 OpenMP threads)

Once the 40 physical cores threshold is exceeded, the overall performance of the benchmarks is reduced. In the multiple of 10 case, performance degrades even before the maximum thread count is reached.

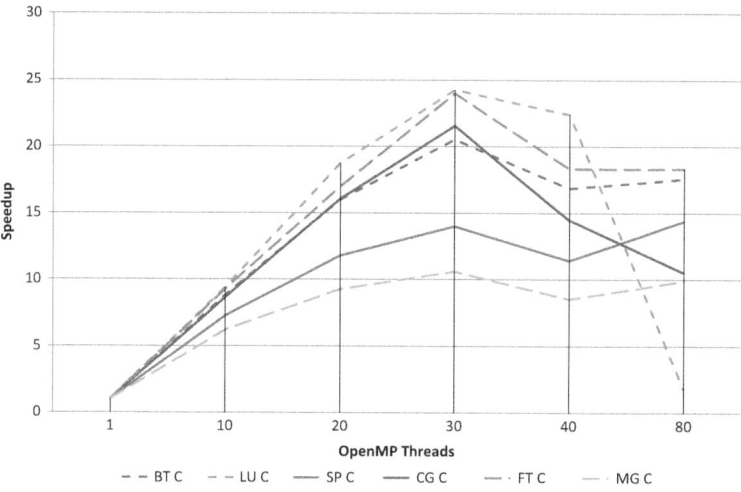

Fig. 33 SuperMUC fat node scaling, NPB MZ benchmarks (multiples of 10 OpenMP threads)

6.4 Performance Summary

In general, the results of this section show that performance with OpenMP is hardware platform specific as well as application specific. Automatic tuners can be of benefit to users and system administrators when migrating to new systems, even with well optimized and mature code bases.

Execution time results were presented in this section. While time is still an essential metric in HPC, energy consumption has been gaining importance in recent years. Looking at the scalability plots, it is apparent that certain thread changes bring minimal benefits to the performance metric. It is under such cases that multi-objective optimizations can bring forth near peak performance with significant reductions on energy consumption. For example, in Fig. 31 it is shown that the performance of the multi-grid solver does not scale as strongly as the other benchmarks past the 4 thread threshold. This can motivate us to limit this application to 1 socket with 4 threads, and use the available power budget for higher frequencies.

The results were collected with a single pinning strategy that maximizes the bandwidth available to each thread. Most HPC workloads benefit from increased bandwidth, so this is a sensible default. Nevertheless, thread pinning can affect performance significantly, and is a worthwhile parameter to tune as well, particularly when dealing with core counts that are lower than the available hardware threads in NUMA systems.

It can be concluded that automatic tuning techniques are already valuable for tuning single aspects of parallel applications. Their utility can only increase, as different aspects and tunable parameters are added to the search spaces considered and manual evaluation processes become increasingly unmanageable for users.

The Periscope Tuning Framework (PTF) provides automatic tuning plugins for common programming models and use cases, as well as an infrastructure for quick development of custom ones. Applications that do multi-threading through OpenMP are already supported by the framework, through the OpenMP scalability plugin evaluated in this section.

7   Conclusions

This article presented the Periscope Tuning Framework (PTF) that was developed in the European AutoTune project. It supports automatic tuning for HPC applications via tuning plugins. Tuning plugins are aspect specific and exploit expert knowledge to handle big search spaces.

Several plugins were developed for different programming models and programming aspects in the project. PTF is designed to be extensible with new tuning plugins. We gave an introduction into how to program such new tuning plugins. We presented the general structure of a plugin as well as how it can be extended. Examples were given for features such as pre-analysis to guide the tuning plugin in exploring the tuning space, the usage of analysis services during the experiment with a certain tuning scenario and the parallel evaluation of scenarios if multiple MPI processes are available.

Additional information on how to write tuning plugins in more detail is available at the Periscope home page (http://periscope.in.tum.de) as well as in upcoming tutorials.

References

1. Auweter, A., Bode, A., Brehm, M., Brochard, L., Hammer, N., Huber, H., Panda, R., Thomas, F., Wilde, T.: A case study of energy aware scheduling on supermuc. In: International Supercomputing Conference (ISC) Proceedings 2014, Leipzig (2014)
2. Benkner, S., Pllana, S., Traff, J., Tsigas, P., Dolinsky, U., Augonnet, C., Bachmayer, B., Kessler, C., Moloney, D., Osipov, V.: Peppher: efficient and productive usage of hybrid computing systems. IEEE Micro 31(5), 28–41 (2011)
3. Brunst, H., Hackenberg, D., Juckeland, G., Rohling, H.: Comprehensive peformance tracking with Vampir 7. In: Müller, M., Resch, M., Schulz, A., Nagel, W. (eds.) Tools for High Performance Computing, pp. 17–30. Springer, Heidelberg/London (2010)
4. Geimer, M., Wolf, F., Wylie, B., Becker, D., Böhme, D., Frings, W., Hermanns, M., Mohr, B., Szebenyi, Z.: Recent developments in the scalasca toolset. In: Müller, M., Resch, M., Schulz, A., Nagel, W. (eds.) Tools for High Performance Computing, pp. 39–52. Springer, Heidelberg/London (2010)
5. GNU Coding Standards: http://www.gnu.org/prep/standards/standards.html (2014)
6. Kukkonen, S., Lampinen, J.: Gde3: the third evolution step of generalized differential evolution. In: The 2005 IEEE Congress on Evolutionary Computation, Edinburgh, vol. 1, pp. 443–450. IEEE (2005)

7. Score-P: Scalable Performance Measurement Infrastructure for Parallel Codes. http://www.vi-hps.org/projects/score-p/ (2015)
8. Shende, S.S., Malony, A.D.: The TAU parallel performance system. Int. J. High Perform. Comput. Appl. ACTS Collection Special Issue **20**(2), 287–311 (2006)
9. Tiwari, A., Hollingsworth, J.: Online adaptive code generation and tuning. In: 2011 IEEE International Parallel Distributed Processing Symposium (IPDPS), Anchorage, May 2011, pp. 879–892 (2011)

Combining Instrumentation and Sampling for Trace-Based Application Performance Analysis

Thomas Ilsche, Joseph Schuchart, Robert Schöne, and Daniel Hackenberg

Abstract Performance analysis is vital for optimizing the execution of high performance computing applications. Today different techniques for gathering, processing, and analyzing application performance data exist. Application level instrumentation for example is a powerful method that provides detailed insight into an application's behavior. However, it is difficult to predict the instrumentation-induced perturbation as it largely depends on the application and its input data. Thus, sampling is a viable alternative to instrumentation for gathering information about the execution of an application by recording its state at regular intervals. This method provides a statistical overview of the application execution and its overhead is more predictable than with instrumentation. Taking into account the specifics of these techniques, this paper makes the following contributions: (I) A comprehensive overview of existing techniques for application performance analysis. (II) A novel tracing approach that combines instrumentation and sampling to offer the benefits of complete information where needed with reduced perturbation. We provide examples using selected instrumentation and sampling methods to detail the advantage of such mixed information and discuss arising challenges and prospects of this approach.

1 Introduction

Performance analysis tools allow users to gain insight into the run-time behavior of applications and improve the efficient utilization of computational resources. Especially for complex parallel applications, the concurrent behavior of multiple tasks is not always obvious, which makes the analysis of communication and synchronization primitives crucial to identify and eliminate performance bottlenecks.

T. Ilsche (✉) • J. Schuchart • R. Schöne • D. Hackenberg
Center for Information Services and High Performance Computing (ZIH), Technische
Universität Dresden, 01062 Dresden, Germany
e-mail: thomas.ilsche@tu-dresden.de; joseph.schuchart@tu-dresden.de;
robert.schoene@tu-dresden.de; daniel.hackenberg@tu-dresden.de

© Springer International Publishing Switzerland 2015
C. Niethammer et al. (eds.), *Tools for High Performance Computing 2014*,
DOI 10.1007/978-3-319-16012-2_6

Different techniques for conducting performance analyses have been established, each with their specific set of distinct advantages and shortcomings. These techniques differ in the type and amount of information they provide, e.g., about the behavior of one process or thread and the interaction between these parallel entities, the amount of data that is generated and stored, as well as the level of detail that is contained within the data. One contribution of this paper is to give a structured overview on these techniques to help users understand their nature. However, most of these approaches suffer from significant peculiarities or even profound disadvantages that limit their applicability for real-life performance optimization tasks:

- Full application instrumentation provides exhaustive information but comes with unpredictable program perturbation that can easily conceal the performance characteristics that need to be analyzed. Extensive event filtering may reduce the overhead, but this does require additional effort.
- Pure MPI instrumentation mostly comes with low overhead, but it provides only very limited information as the lack of application context for communication patterns complicates the performance analysis and optimization.
- Pure sampling approaches create very predictable program perturbation, but they lack communication and I/O information. Moreover, the classical combination with profiling for performance data presentation squanders important temporal correlations.
- Instrumentation-based approaches can only access performance counters at application events, thereby hiding potentially important information from in between these events.

A combination of techniques can often leverage the combined advantages and mitigate the weaknesses of individual approaches. We present such a combined approach that features low overhead and a high level of detail to significantly improve the usability and effectiveness of the performance analysis process.

2 Performance Analysis Techniques: Classification and Related Work

The process of performance analysis can be divided into three general steps: data acquisition, data recording, and data presentation [10]. These steps as well as common techniques for each step are depicted in Fig. 1. Data acquisition reveals relevant performance information of the application execution for further processing and recording. This information is aggregated for storage in memory or persistent media in the data recording layer. The data presentation layer defines how the information is presented to the user to create insight for further optimization. In this section we present an overview of the often ambiguously used terminology and the state of the art of performance analysis tools.

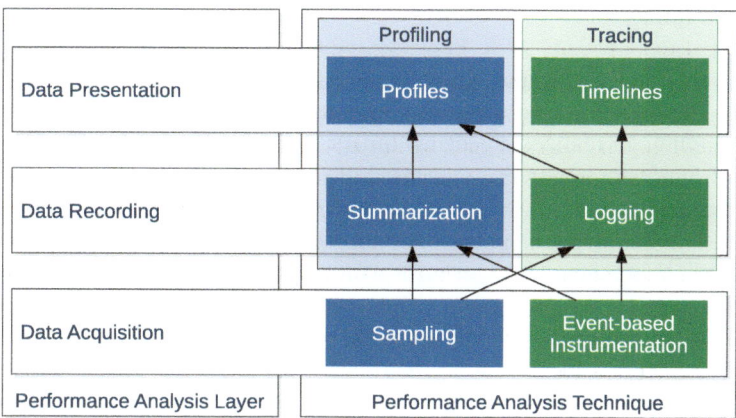

Fig. 1 Classification of performance analysis techniques (based on [11]). Valid combinations of techniques are connected with an *arrow*. Presenting data recorded by logging as a profile requires a post-processing summarization step

2.1 Data Acquisition

2.1.1 Event-Based Instrumentation

Event-based instrumentation refers to a modification of the application execution in order to record and present certain intrinsic events of the execution, e.g., function entry and exit events. After the modification, these events trigger the data recording by the measurement environment at run-time. More specific events with additional semantics, such as communication or I/O operations, can often be derived from the execution of an API function.

The modification of the application can be applied on different levels. *Source code instrumentation* APIs used for a manual instrumentation, *source-to-source transformation* tools like PDT [14] and Opari [16], and *compiler instrumentation* require analysts to recompile the application under investigation after inserting instrumentation points manually or automatically. Thus, they can only be used for applications whose source code is available. Common ways to instrument applications without recompilation are *library wrapping* [5], *binary rewriting* (e.g., via DYNINST [3] or PEBIL [13]), and *virtual machines* [2].

All of these techniques are often referred to as *event-based instrumentation*, *direct instrumentation* [23], *event trigger* [11], *probe-based* measurement [17] or simply *instrumentation* and it is common to combine several of them in order to gather information on different aspects of an application run.

2.1.2 Sampling

Another common technique to obtain performance data is *sampling*, which describes the periodic interruption of a running program and inspection of its state. Sampling is realized by using timers (e.g., `setitimer`) or an overflow trigger of hardware counters (e.g., using PAPI [6]). The most important aspects of inspecting the state of execution are the call-path and hardware performance counters. The call-path provides information about all functions (and regions) that are currently being executed. This information roughly corresponds to the enter/exit function events from event-based instrumentation. Additionally, the instruction pointer can be obtained, allowing sampling to narrow down hot-spots even within functions. However, the semantic interpretation of specific API calls is limited and can prevent the reconstruction of process interaction or I/O due to missing information. Moreover, the state of the application between two sampling points is unavailable for analysis.

In contrast to event-based instrumentation, sampling has a much more predictable overhead that mainly depends on the sampling rate rather than the event frequency. The user specifies the sampling rate and thereby controls the trade-off between measurement accuracy and overhead. While the complete information on specific events is not guaranteed with sampling, the recorded data can provide a statistical basis for analysis. For this reason, *sampling* is sometimes also referred to as *statistical sampling* or *profiling*.

2.2 Data Recording

2.2.1 Logging

Logging is the most elaborate technique for recording performance data. A timestamp is added to the information from the acquisition layer and all the information is retained in the recorded data. It can apply to both data from sampling and event-based instrumentation. Logging requires a substantial amount of memory and can cause perturbation and overhead during the measurement due to the I/O operations for writing a log-file to persistent storage. The term *tracing* is often used synonymously to *logging* and the data created by *logging* is a *trace*.

2.2.2 Summarization

By *summarizing* the information from the acquisition layer, the memory requirements and overhead of data recording are minimized at the cost of discarding the temporal context. For event-based instrumentation, values like sum of event duration, event count, or average message size can be recorded. Summarization of samples mainly involves counting how often a specific function is on the call-path, but performance metrics can also be summarized. This technique is also called *profiling*, because the data presentation of a summarized recording is a profile. A special hybrid case is the *phase* profile [15] or *time-series* profile [24], for which the information is summarized separately for successive phases (e.g., iterations) of the application. This provides some insight into the temporal behavior, but not to the extent of logging.

2.3 Data Presentation

2.3.1 Timelines

A timeline is a visual display of an application execution over time and represents the temporal relationship between events of a single or different parallel entities. This gives a detailed understanding of how the application is executed on a specific machine. In addition to the time dimension, the second dimension of the display can depict the call-path, parallel execution, or metric values. An example is given in Fig. 2. Necessarily, timelines can only be created from logged data, not from summarized data.

2.3.2 Profiles

In a profile, the performance metrics are presented in a summary that is grouped by a factor such as the name of the function (or region). A typical profile is provided in Listing 1 and shows the distribution of the majority of time spent among functions. In such a *flat* profile the information is grouped by function name. It is also possible to group the information based on the call-path resulting in a *call-path* profile [24] (or *call graph* profile [8]). For performance metrics, the grouping can be done by metric or a combination of call-path and metric. Profiles can be created from either summarized data or logs.

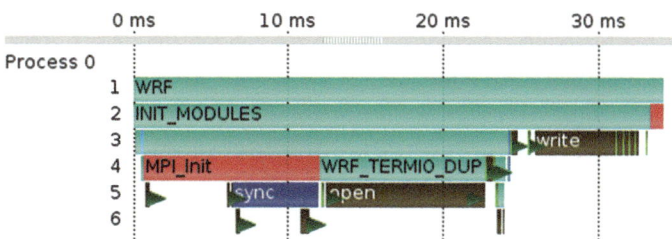

Fig. 2 A process timeline displaying the call-path and event annotations

```
Each sample counts as 0.01 seconds.
%    cumulative   self              self    total
time   seconds   seconds calls ms/call  ms/call name
33.34    0.02      0.02   7208   0.00      0.00   open
16.67    0.03      0.01    244   0.04      0.12   offtime
16.67    0.04      0.01      8   1.25      1.25   memccpy
16.67    0.05      0.01      7   1.43      1.43    write
```

Listing 1 Example output of gprof taken from its manual [19]

2.4 Event Types

2.4.1 Code Regions

Several event types are of interest for application analysis. By far the most commonly used event types are code regions, which can be function calls either inside the application code or to a specific library, or more generally be any type of region such as loop bodies and other code structures. Therefore, code regions within the application are in the focus of this work. The knowledge of the execution time of an application function and its corresponding call-path is imperative for the analysis of application behavior. However, function calls can be extremely frequent and thus yield a high rate of trace events. This is especially true for C++ applications, where short methods are very common, making it difficult to keep the run-time overhead of instrumentation and tracing low.

2.4.2 Communication and I/O Operations

The exchange of data between tasks (communication) is essential for parallel applications and highly influential on the overall performance. Communication events can contain information about the sender/receiver, message size, and further context such as MPI tags. File I/O is a form of data transfer between a task and persistent storage. It is another important aspect for application performance. Typical file I/O events include information about the active task, direction (read/write), size, and file name.

2.4.3 Performance Metrics

The recording of the above mentioned events only gives limited information on the usage efficiency of shared and exclusive resources. Additional metrics describing the utilization of these resources are therefore important performance measures. The set of metrics consists of (but is not limited to) hardware performance counter (as provided by PAPI), operating system metrics (e.g., via rusage), and energy and power measurements.

2.4.4   Task Management

The management of tasks (processes and threads) is also of interest for application developers. This set of events includes task creation (fork), shutdown (join), and the mapping from application tasks to OS threads.

2.5   *Established Performance Analysis Tools*

Several tools support the different techniques mentioned in Sect. 2 and in parts combine some of them.

The Scalasca [7] package focuses on displaying profiles, but logged data is used for a special post-processing analysis step. VampirTrace [18] mainly focuses on refined tracing techniques but comes with a basic profiling mode and external tools for extracting profile information from trace data. These two software packages rely mostly on different methods of event-based instrumentation. The Tuning and Analysis Utilities (TAU) [22] implement a measurement system specialized for profiling with some functionality for tracing. TAU supports a wide range of instrumentation methods but a hybrid mode that uses call-path sampling in combination with instrumentation is also possible [17]. The performance measurement infrastructure Score-P [12] has both sophisticated tracing and profiling capabilities. It mainly acquires data from event-based instrumentation, but recent work [23] introduced call-path sampling for profiling. The graphical tool Vampir [18] can visualize traces created with Score-P, VampirTrace or TAU in the form of timelines or profiles. Similar to the above mentioned, the Extrae software records traces based on various instrumentation mechanisms. Sampling in Extrae is supported by interval timers and hardware performance counter overflow triggers. The sampling data of multiple executions of a single code region can be combined into a single detailed view using folding [21]. This combined approach provides increased information about repetitive code regions. HPCToolkit [1] implements sampling based performance recording. It provides sophisticated techniques for stack unwinding and call-path profiling. The data can also be recorded in a trace and displayed in a timeline trace viewer. All previously mentioned tools have a strong HPC background and are therefore designed to analyze large scale programs. For example Scalasca and VampirTrace/Vampir can handle applications running on more than 200,000 cores [9, 25].

Similar combinations of techniques can also be seen in tools without a specialization for HPC. The Linux' perf infrastructure [4] consists of a user space tool and a kernel part that allows for application-specific and system-wide sampling based on both hardware events and events related to the operating system itself. Support for instrumentation-based analysis is added through kprobes, uprobes, and tracepoint events. The infrastructure part of perf is also used by many other tools as it provides the basis to read hardware performance counters on Linux with PAPI. The GNU profiler (gprof) [8] provides a statistical profile of function run-times, but also employs instrumentation by the compiler to derive accurate number-of-calls figures.

3 Combining Multiple Performance Analysis Techniques: Concept and Experiences

As discussed in Sect. 2, sampling and event-based instrumentation have different strengths and weaknesses. A combined performance analysis approach can use instrumentation for aspects of the application execution for which full information is desired and sampling to complement the performance information with limited perturbation. We discuss two new approaches and evaluate them based on prototype implementations for the VampirTrace plugin counter interface [20]: (I) Instrumenting MPI calls and sampling call-paths; and (II) Instrumenting application regions but sampling hardware performance counters.

3.1 MPI Instrumentation and Call-Path Sampling

Performance analysis of parallel applications is often centered around messages and synchronization between processes. In the case of applications using MPI, it is common practice to instrument the API calls to get information about every message during application execution [7, 15, 18, 22]. The MPI profiling interface (PMPI) allows for a convenient and reliable instrumentation that only requires re-linking and can even be done dynamically when using shared libraries. Using sampling for message passing information would significantly limit the analysis, e.g., since reliable message matching requires information about each message. However, only recording message events lacks context for a holistic analysis, as for example the root cause of inefficient communication or load imbalances cannot be determined. Call-path sampling is a viable option to complement message recording, as it provides rich context information but – unlike compiler instrumentation – does not require recompilation. The projected run-time perturbation and overhead of this approach is very promising: On the one hand, the overhead can be controlled by adjusting the sampling rate. On the other hand, MPI calls for communication can be assumed to have a certain minimum run-time, thereby limiting the event frequency as well as the overhead caused by this instrumentation. Some applications that make excessive use of many small messages, especially when using non-blocking MPI functions, are still difficult to analyze efficiently with this approach, but this also applies to MPI only instrumentation.

3.1.1 Implementation

We implemented a prototypical sampling support for VampirTrace as a plugin. Whenever VampirTrace registers a task for performance analysis, the plugin is activated and initializes a performance counter based interrupt, e.g., every 1 million cycles. Whenever such a counter overflow occurs, the plugin checks whether the current functions on the stack belong to the main application, i.e., are not part of a library, and adds function events for all functions on the call-path. MPI library calls and communication events are recorded using the instrumented MPI library of VampirTrace. The application does not have to be recompiled to create a trace.

3.1.2 Results

Figure 3 shows the visualization of a trace using an unmodified version of Vampir [18], i.e., without specific support for sampled events. The MPI function calls and messages are clearly visible due to the instrumented MPI library. The application functions, and thus the context of the communication operation, are visible as samples. This already allows users to analyze the communication, possible bottlenecks, and imbalances. Containing the complete call stack in the trace remains as future work.

Figure 4 shows the measured overhead for recording traces of the analyzed NPB benchmark. The overhead is very high for the fully instrumented version, while sampling application functions in addition to the instrumented MPI library only adds a marginal overhead. Thus, while providing all necessary information

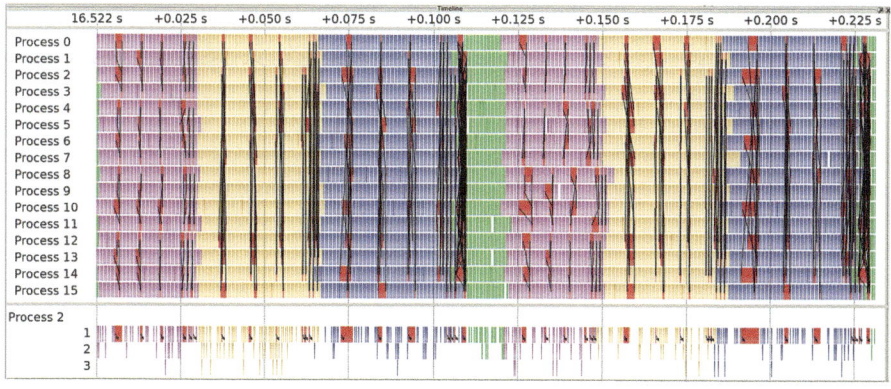

Fig. 3 Vampir visualization of a trace of the NPB BT MPI benchmark created using an instrumented MPI library (MPI functions displayed *red* and messages as *black lines*) and sampling for application functions (x_solve colored *pink*, y_solve *yellow*, z_solve *blue*). Stack view of one process shown below the master timeline

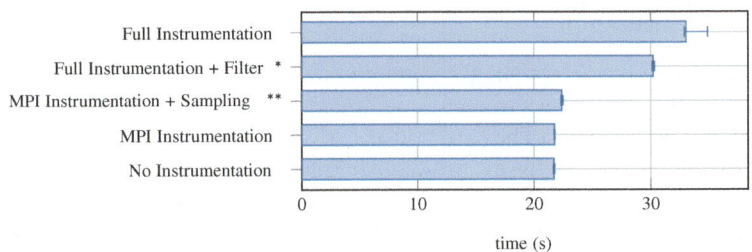

Fig. 4 Run-time of different performance measurement methods for NPB BT CLASS B, SIZE 16 on a dual socket Sandy Bridge system. Median of 10 repeated runs with minimum/maximum bars. * Filtered functions: `matmul_sub`, `matvec_sub`, `binvrhs`, `binvcrhs`, `lhsinit`, `exact_solution`; ** Sampling rate of 2.6 kSa/s

on communication events and still allowing the analysis of the application's call-paths, the overhead can be decreased significantly. These results demonstrate the advantage of combining call-path sampling and library instrumentation.

3.2 Sampling Hardware Counters and Instrumenting Function Calls and MPI Messages

As a second example, we demonstrate the sampling of hardware counter values while tracing function calls and MPI events with traditional instrumentation. In contrast to the traditional approach of recording hardware counter values on every application event, this approach has two important advantages: First, in long running code regions with filtered or no subroutine calls, the sampling approach still provides intermediate data points that allow users to estimate the application performance for smaller parts of this region. Second, for very short code regions, the overhead of the traditional approach can cause significant program perturbation and recorded performance data that does not necessarily contain valuable information for the optimization process. Moreover, reading hardware counter values in short running functions can cause misleading results due to measurement perturbation.

3.2.1 Implementation

For each application thread, the plugin creates a monitoring thread that wakes up in certain intervals to query and record the hardware counters and sleeps the rest of the time.

3.2.2 Results

Figure 5 shows the visualization of a trace of NPB FT that was acquired using compiler instrumentation and an instrumented MPI library. The trace contains two

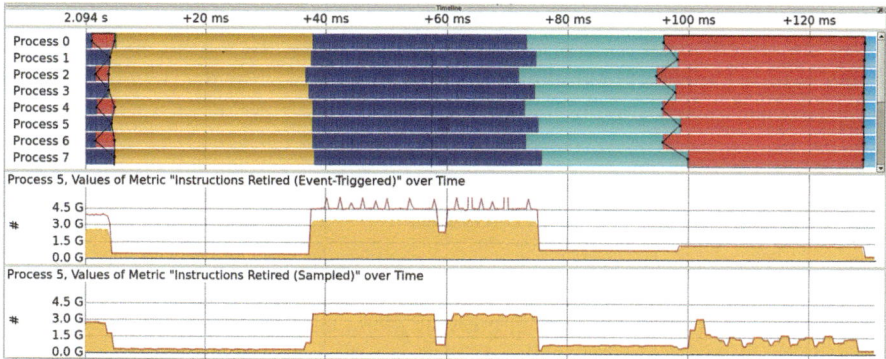

Fig. 5 Vampir visualization of a trace of the NPB FT benchmark acquired through compiler instrumentation and instrumented MPI library (master timeline, *top*) including an event-triggered (*middle*) and a sampled (*bottom*) counter for retired instructions. Colors: MPI *red*, FFT *blue*, evolve *yellow*, transpose *light blue*

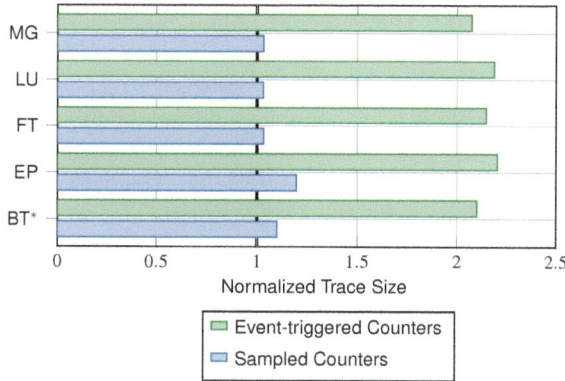

Fig. 6 Normalized trace sizes of NPB CLASS B benchmarks containing hardware performance counters either triggered by instrumentation events or asynchronously sampled (1 kSa/s). Baseline: trace without counters. * Filtered functions: matmul_sub, matvec_sub, binvcrhs, exact_solution

different versions of the same counter (retired instructions), one recorded on every enter/exit event (middle part) and the second sampled every 1 ms (bottom). On the one hand, the instrumented counter shows peaks in regions with a high event rate due to very short-running functions. This large amount of information is usually of limited use except for analyzing these specific function calls. The sampled counter does not provide this wealth of information but still reflects the average application performance in these regions correctly. On the other hand, the sampled counter provides additional information for long running regions, e.g., MPI functions and the evolve_ function. This information is useful for having a more fine-grained estimation of the hardware resource usage of these code areas. Furthermore, Fig. 6

demonstrates that sampling counter values can be used to significantly reduce trace sizes compared to recording counter values through instrumentation. After all, combining the approaches outlined in this section and in Sect. 3.1 is feasible and will remain as future work.

4 Conclusions and Future Work

In this paper, we presented a comprehensive overview of existing performance analysis techniques and the tools employing them, taking into account their specific advantages and disadvantages. In addition, we discussed the general approach of combining the existing techniques of instrumentation and sampling to leverage each of their potential. We demonstrated this with two practical examples, showing results of prototype implementations for (I) sampling application function call-paths while instrumenting MPI library calls; and (II) sampling hardware performance counter values in addition to traditional application instrumentation. The results confirm that this combined approach has unique advantages over the individual techniques.

Based on the work presented here, we will continue to explore ways of combining instrumentation and sampling for performance analysis by integrating and extending open-source tools available for both strategies. Taking more event types into consideration is another important aspect. For instance, I/O operations and CUDA API calls are viable targets for instrumentation while resource usage (e.g. memory) can be sampled.

Another interesting aspect is the visualization of traces based on call-path samples in a close-up view. It is challenging to present this non-continuous information in an intuitively understandable fashion. We will also further investigate the scalability of our combined approach. The effects of asynchronously sampling in large scale systems that require a very low OS noise to operate efficiently needs to be studied. Our goal is a seamless integration of instrumentation and sampling for gathering trace data to be used in a scalable and holistic performance analysis technique.

Acknowledgements This work has been funded by the Bundesministerium für Bildung und Forschung via the research project CoolSilicon (BMBF 16N10186) and the Deutsche Forschungsgemeinschaft (DFG) via the Collaborative Research Center 912 "Highly Adaptive Energy-Efficient Computing" (HAEC, SFB 921/1 2011). The authors would like to thank Michael Werner for his support.

References

1. Adhianto, L., Banerjee, S., Fagan, M., Krentel, M., Marin, G., Mellor-Crummey, J., Tallent, N.R.: HPCTOOLKIT: tools for performance analysis of optimized parallel programs. Concurr. Comput.: Pract. Exp. **22**(6), 685–701 (2010)

2. Binder, W.: Portable and accurate sampling profiling for Java. Softw.: Pract. Exp. **36**(6), 615–650 (2006)
3. Buck, B., Hollingsworth, J.K.: An API for runtime code patching. Int. J. High Perform. Comput. Appl. **14**, 317–329 (2000)
4. de Melo, A.C.: The new linux 'perf' tools. In: Slides from Linux Kongress, The German Unix User Group (2010)
5. Dietrich, R., Ilsche, T., Juckeland, G.: Non-intrusive performance analysis of parallel hardware accelerated applications on hybrid architectures. In: International Conference on Parallel Processing Workshops, San Diego (2010)
6. Dongarra, J., Malony, A.D., Moore, S., Mucci, P., Shende, S.: Performance instrumentation and measurement for terascale systems. In: Proceedings of the 2003 International Conference on Computational Science, ICCS'03, Melbourne. Springer (2003)
7. Geimer, M., Wolf, F., Wylie, B.J.N., Ábrahám, E., Becker, D., Mohr, B.: The Scalasca performance toolset architecture. Concurr. Comput.: Pract. Exp. **22**(6), 702–719 (2010)
8. Graham, S.L., Kessler, P.B., McKusick, M.K.: gprof: a call graph execution profiler. In: SIGPLAN Symposium on Compiler Construction, Boston (1982)
9. Ilsche, T., Schuchart, J., Cope, J., Kimpe, D., Jones, T., Knüpfer, A., Iskra, K., Ross, R., Nagel, W.E., Poole, S.: Optimizing I/O forwarding techniques for extreme-scale event tracing. Cluster Comput. **9**, 1–18 (2013)
10. Jain, R.K.: The Art of Computer Systems Performance Analysis: Techniques for Experimental Design, Measurement, Simulation, and Modeling. Wiley, New York (1991)
11. Juckeland, G.: Trace-based performance analysis for hardware accelerators. PhD thesis, TU Dresden (2012)
12. Knüpfer, A., Rössel, C., an Mey, D., Biersdorff, S., Diethelm, K., Eschweiler, D., Geimer, M., Gerndt, M., Lorenz, D., Malony, A.D., Nagel, W.E., Oleynik, Y., Philippen, P., Saviankou, P., Schmidl, D., Shende, S.S., Tschüter, R., Wagner, M., Wesarg, B., Wolf, F.: Score-P – a joint performance measurement run-time infrastructure for Periscope, Scalasca, TAU, and Vampir. In: Proceedings of 5th Parallel Tools Workshop, 2011, Dresden. Springer (2012)
13. Laurenzano, M.A., Tikir, M.M., Carrington, L., Snavely, A.: Pebil: efficient static binary instrumentation for linux. In: IEEE International Symposium on Performance Analysis of Systems Software (ISPASS), White Plains (2010)
14. Lindlan, K.A., Cuny, J., Malony, A.D., Shende, S., Juelich, F., Rivenburgh, R., Rasmussen, C., Mohr, B.: A tool framework for static and dynamic analysis of object-oriented software with templates. In: Proceedings of the International Conference on Supercomputing, Santa Fe. IEEE (2000)
15. Malony, A.D., Shende, S.S., Morris, A., Joubert, G.R., Nagel, W.E., Peters, F.J., Plata, O., Tirado, P., Zapata, E.: Phase-based parallel performance profiling. In: Proceedings of the PARCO 2005 Conference, jülich, Malaga (2005)
16. Mohr, B., Malony, A.D., Shende, S., Wolf, F.: Towards a performance tool interface for OpenMP: An approach based on directive rewriting. In: Proceedings to the Third Workshop on OpenMP (EWOMP), Barcelona (2001)
17. Morris, A., Malony, A.D., Shende, S., Huck, K.A.: Design and implementation of a hybrid parallel performance measurement system. In: ICPP, San Diego, pp. 492–501 (2010)
18. Müller, M.S., Knüpfer, A., Jurenz, M., Lieber, M., Brunst, H., Mix, H., Nagel, W.E.: Developing scalable applications with Vampir, VampirServer and VampirTrace. In: Parallel Computing: Architectures, Algorithms and Applications, vol. 15. IOS Press, Amsterdam/Washington, DC (2008)
19. Osier, J.: The GNU gprof manual (2014)
20. Schöne, R., Tschüter, R., Ilsche, T., Hackenberg, D.: The vampirtrace plugin counter interface: introduction and examples. In: Euro-Par 2010 Parallel Processing Workshops, Ischia. Volume 6586 of Lecture Notes in Computer Science. Springer (2011)
21. Servat, H., Llort, G., Giménez, J., Huck, K., Labarta, J.: Folding: detailed analysis with coarse sampling. In: Tools for High Performance Computing 2011, Dresden. Springer (2012)

22. Shende, S.S., Malony, A.D.: The TAU parallel performance system. Int. J. High Perform. Comput. Appl. **20**(2), 287–311 (2006)
23. Szebenyi, Z., Gamblin, T., Schulz, M., de Supinski, B.R., Wolf, F., Wylie, B.J.N.: Reconciling sampling and direct instrumentation for unintrusive call-path profiling of MPI programs. In: Proceedings of the 25th IEEE International Parallel & Distributed Processing Symposium (IPDPS), Anchorage, May 2011
24. Szebenyi, Z., Wolf, F., Wylie, B.J.N.: Space-efficient time-series call-path profiling of parallel applications. In: Proceedings of the International Conference on Supercomputing, Yorktown Heights, Nov 2009. ACM (2009)
25. Wylie, B.J.N., Geimer, M., Mohr, B., Böhme, D., Szebenyi, Z., Wolf, F.: Large-scale performance analysis of Sweep3D with the Scalasca toolset. Parallel Process. Lett. **20**(4), 397–414 (2010)

Ocelotl: Large Trace Overviews Based on Multidimensional Data Aggregation

Damien Dosimont, Youenn Corre, Lucas Mello Schnorr, Guillaume Huard, and Jean-Marc Vincent

Abstract Performance analysis of parallel applications is commonly based on execution traces that might be investigated through visualization techniques. The weak scalability of such techniques appears when traces get larger both in time (many events registered) and space (many processing elements), a very common situation for current large-scale HPC applications. In this paper we present an approach to tackle such scenarios in order to give a correct overview of the behavior registered in very large traces. Two configurable and controlled aggregation-based techniques are presented: one based exclusively on the temporal aggregation, and another that consists in a spatiotemporal aggregation algorithm. The paper also details the implementation and evaluation of these techniques in **Ocelotl**, a performance analysis and visualization tool that overcomes the current graphical and interpretation limitations by providing a concise overview registered on traces. The experimental results show that Ocelotl helps in detecting quickly and accurately anomalies in 8 GB traces containing up to 200 million of events.

1 Introduction

Performance analysts commonly use execution traces collected at runtime to understand the behavior of applications running on distributed and parallel systems. These traces are then inspected post mortem using various interactive visualization techniques. They are used by performance analysts to detect unexpected behavior, correlating the events to the application source code and its components, execution settings or the configuration of the runtime platform. A very common example of visualization technique is the Gantt-like space/time visualization, which represents

D. Dosimont (✉) • Y. Corre • G. Huard • J.-M. Vincent
Inria, University of Grenoble Alpes, LIG, F-38000 Grenoble, France

CNRS, LIG, F-38000 Grenoble, France
e-mail: damien.dosimont@imag.fr; youenn.corre@inria.fr; guillaume.huard@imag.fr;
jean-marc.vincent@imag.fr

L.M. Schnorr
Informatics Institute, UFRGS, Porto Alegre, Brazil
e-mail: schnorr@inf.ufrgs.br

© Springer International Publishing Switzerland 2015
C. Niethammer et al. (eds.), *Tools for High Performance Computing 2014*,
DOI 10.1007/978-3-319-16012-2_7

the behavior of the application components along a time axis. The problem is that such visualization technique is incapable to scale properly for an increasing number of events. This issue is mainly due to the bounded screen resolutions, preventing the proper drawing of many graphical objects (see Fig. 1 for an example). Another problem is the natural limitation of human perception, which cannot handle too much irregular information. From the technical side, long processing time and interaction delay, caused by the trace event handling, may hinder and discourage the analyst to conduct further behavior investigation.

We present **Ocelotl**, a tool that overcomes the graphical and interpretation limitations of current visualization techniques. It provides a concise overview of the trace behavior, as the result of a user-configurable and controlled data aggregation process. The aggregation process gathers parts of the trace where behavior is homogeneous. Such homogeneity is determined by a user-configurable trade-off between the information lost by the aggregation process, and the complexity reduction of the visual representation. The user dynamically adjusts the balance between both metrics to get a level of details that helps him understand the application behavior.

We propose two different trace visualization overviews based on this trade-off principle: a temporal and a spatiotemporal-based approach. In the former, the aggregation is applied exclusively on the time dimension. The obtained aggregated view is mainly aimed at analyzing applications expected to be regular along time. Examples of such applications are the software of multimedia players or algebraic computation: we easily highlight temporal perturbations and phases on these use-cases. The later, spatiotemporal, is when the aggregation is mutually applied in time and space dimensions. By space we mean the hardware and software component hierarchy. This second technique is efficient to detect problematic behaviors that touch only a subset of resources, which is typical of network sharing issues or unbalanced performances of machines and clusters running a parallel application.

Because of the computational cost derived from the efficient implementation of these techniques, we have designed a clever reuse of results provided by previous analysis sessions. It helps to drastically reduce the computation time and improve the interaction responsiveness and the general utilization of the tool. The experimental results show that Ocelotl helps to quickly and accurately detect anomalies in 8 GB traces containing up to 200 million of events.

This text is organized as follows. Section 2 presents the state-of-the-art on performance analysis visualization tools, with a focus on the relationship with visual and semantic aggregation mechanisms that are present on related work. Section 3 reminds two aggregation methodologies based on a trade-off between information loss and complexity reduction that were described in previous documents. Section 4 presents the Ocelotl performance visualization tool that implements the overviewing techniques of our approach. Section 5 closes the paper with a work summary and future work.

Fig. 1 Example of a cluttered space/time view that is incapable to show correctly a very large trace over temporal and spatial dimensions

2   Trace Overview Common Issues

Different analysis tools propose several methods to improve scalability and provide decent overviews over time and space. In this section, we describe several common issues related to the techniques they employ.

2.1   Pixel Guided Representations

Pixel-guided representations, present in some tools, like Vampir or Paraver [5, 14], associate each screen pixel to a set of data. As the pixel is incapable of representing all the information, the rendering algorithm decides what is shown or hidden. We claim that this aggregation process misguides the user: a pixel might correspond to a raw data or be the aggregation of several data. Moreover, some tools do not provide clear indication about the aggregation operation used to render the pixel. Pixel-guided representations also suffer from a fidelity issue: for instance, resizing the window may modify strongly the visualization content because the pixel allocation changes.

2.2   Visual Aggregation

In a visual aggregation, the rendering tries to preserve the graphical object scales, whose size depends on their contents. When it is impossible, e.g. when the size is less than one pixel, it generates aggregates gathering close objects. In Pajé [1] and LTTng Eclipse Viewer [11], such aggregates are just used to avoid visual clutter, but do not represent the data they contain.

2.3   Spatiotemporal Entity Budget Management

Techniques based on Gantt Chart-like space-time diagrams [1, 5, 11, 12, 14, 15] have an issue to keep a reasonable budget of graphical entities present on the screen. Indeed, since the temporal and the spatial dimensions behave differently, visualization techniques do not treat both axis the same way. KPTrace Viewer [15], for instance, proposes an interactive hierarchical aggregation for the space dimension and a mechanism that compresses time whenever an event lasts too long. However, if the spatial reduction technique is efficient, the time compression does not manage sufficiently the time entity budget. Other space-time tools [1, 5, 11, 12, 14] lack reduction technique on space and disrespect the vertical budget, which forces the user to scroll over y-axis and prevents him from getting a spatial overview.

2.4 Monodimensional Representations

Conversely, other techniques [5, 16] propose unidimensional overviews that better manage the entity budget. Vampir's task profile [5] clusters the most similar processes according to a distance measure based on the duration of the functions executed by each process. However, the temporal dimension is lost in the process, and this technique is thus more adapted for profiling. One of our previous works, Viva [10], provides a treemap view showing the hardware and software component hierarchy. Entities having the most homogeneous behavior are aggregated using a compromise between the representation complexity and the information loss induced by the aggregation. This technique highlights troubles characterized by an heterogeneous behavior. However, time dimension is missing from the representation, although it is used to compute the entities values through a time integration. It is thus difficult to find local temporal perturbations, and this technique is thus mainly aimed at finding bottlenecks that concern the whole execution.

3 Provide a Trace Overview Using Data Aggregation

In this section, we remind two applications of an aggregation methodology we already described in previous documents: temporal [3, 4, 13] and spatiotemporal [2] data aggregations that provide overviews of the trace and address issues evoked in Sect. 2. For each technique, we designed an algorithm which gathers the parts of the trace where the behavior is homogeneous. The aggregation strength is tuned thanks to a parameter set by the user, which balances a trade-off between the information lost by the aggregation process and the complexity reduction compared to a representation without aggregation. We also designed visualization techniques to represent the algorithm output. They may involve visual aggregation to help providing a clean representation of the algorithm output. With the set of techniques we use to build our overviews, we reach the objective of reducing the complexity of the representation while keeping important information about the application behavior. We thus address in priority the screen limitations and the analyst understanding issues. Nevertheless, the complexity of our algorithms enables to keep reasonable computation times, and ensures a good interactivity. Both overviews are implemented in the Ocelotl analysis tool, which we describe in Sect. 4.

3.1 Aggregation Methodology

The *analytic approach* consists in analyzing separately each constituent of a system. However, when the system contains too many entities, it is not possible to understand its global behavior from the behavior of each of its components. On the opposite, the *systemic approach* consists in analyzing the system globally, using a macroscopic point of view, in order to provide the analyst with a quantity of information he is able to work with. Lamarche-Perrin proposes an aggregation methodology [7–10], based on the systemic approach, that we follow to build our overview techniques. Since this methodology is generic, we can adapt it to trace analysis.

In order to build the macroscopic overview that represents the behavior of our whole system, we start by observing the system from a microscopic point of view. This observation generates microscopic data, which are, in principle, too complex and too numerous to be analyzed using the analytic approach. Collected information are organized using data structures and metrics which can describe the system behavior. In particular, data are structured over temporal and spatial discrete dimensions.

The microscopic data can be considered as a set of entities x. The aggregation process, which leads to a macroscopic point of view from these microscopic data, is constituted by two steps. First, we partition the entities x, which gives us a partition P of the microscopic model. We name $\mathscr{P}$ the set of all possible partitions P. Then, we generate a result value V_X associated with each part $X \in P$ using an aggregation operator. In order to give meaning to this process, we propose several steps to fulfill:

1. Choose the dimensions where to perform the aggregation. This choice depends on the behavior to study and the visualization that is planned. In this step, we also structure these dimensions (ordered set, hierarchy, etc.).
2. Choose the operand, e.g. the entities that can be aggregated. In our case, it corresponds to the entities x.
3. Constrain the aggregation: determine the set $\mathscr{P}$ of partitions that are allowed. This constraint has several roles. First, it forbids aggregations that cannot translate into an understandable behavior (for instance, in the case of a temporal analysis, we forbid the aggregation of two discontinuous time part). Second, it insures coherence with the structure of the dimensions (for instance, for a hierarchy, we aggregate the nodes into their parents). And last, it reduces the number of possible partitions to limit the aggregation algorithm complexity.
4. Choose the aggregation operator that will generate the aggregation result. This operator can be a mathematical function but also a visual technique.
5. Choose a condition, provided by the user or the environment, which determines how to aggregate the system, i.e. the way the microscopic model is partitioned.

In our methodology, we will use systematically the sum operation to fulfill the step 4: $V_X = \sum_{x \in X} v_x$

3.2 Trigger the Aggregation with Information Loss and Complexity Reduction

A particularity of Lamarche-Perrin methodology is the use of information theory concepts to build the condition triggering the aggregation (step 5). To bring meaning to the analyst, we want to aggregate the microscopic model by gathering in priority the most homogeneous spatiotemporal areas, while preserving data about the heterogeneous ones. In this way, aggregation consists in optimizing a trade-off between data reduction and information loss.

Measures have been proposed in previous work [3, 4, 9, 10, 13] to express such a trade-off. For each part X obtained from a partition P of our microscopic model, we associate measures of information loss (*loss*), data reduction (*gain*) and a *parametrized Information Criterion* (pIC). We use Kullback-Leibler divergence [6] as a measure of information loss:

$$\text{loss}_X = \sum_{x \in X} v_x \log_2 \left(\frac{|X| v_x}{V_X} \right) \tag{1}$$

Shannon entropy [17] as a measure of data reduction,

$$\text{gain}_X = V_X \log_2 V_X - \sum_{x \in X} v_x \log_2 v_x \tag{2}$$

and define the *parametrized Information Criterion* [9] as follow:

$$\text{pIC}_X = p \, \text{gain}_X - (1 - p) \, \text{loss}_X \tag{3}$$

where $p \in [0, 1]$ is the parameter used to balance this trade-off.

These measures are additive. The gain, the loss and the pIC associated with a partition P is defined as the sum of the gain, the loss or the pIC of the parts X that compose it. Finally, we define the optimal partition for a given parameter p as the one that has the highest pIC. This process tends to select partitions that aggregates in priority entities with the closest associated values, since it enables to reduce data complexity while minimizing information loss. For $p = 0$, the analyst wants to be as accurate as possible (the microscopic partition is optimal) and, for $p = 1$, he wants to be the simplest (the full aggregation is optimal). When p varies from 0 to 1, a whole class of nested representations arises. The choice of this parameter is deliberately left to the analyst, so he can adapt the entity budget to the analysis purposes.

3.3   Trace Microscopic Models

In order to apply Lamarche-Perrin methodology to trace analysis, it is mandatory to generate a microscopic model structured by discrete dimensions, and providing a metric that can represent the behavior an analyst wants to analyze. This microscopic model is thus generated from the raw trace and can be considered as a first abstraction of it.

We formalize our microscopic models as matrices involving temporal dimension, spatial dimension, and event type dimension.

- The *time dimension* is defined as the set $T = \{t_1, \ldots, t_n\}$ of the observed microscopic time periods. This dimension is discrete, whereas the raw trace time is continuous. To fit with the microscopic model, the raw trace is divided in $|T|$ (regular) time periods and the events are associated with the periods where they are active. Each period t has a duration $d(t) \in \mathbb{R}^+$ and the whole set is naturally ordered by "the arrow of time". Formally, a total order $<$ on T provides the concept of interval: $T_{(i,j)} = \{t \in T \mid t_i \leq t \leq t_j\}$ with $t_i \leq t_j$. We mark $\mathscr{I}(T)$ the set of intervals of T.
- The *spatial dimension* is defined as the set $S = \{s_1, \ldots, s_m\}$ of the platform microscopic resources. For the computing platforms we are interested in, this set has a hierarchical structure: resources are organized in processes, running on cores, each one being associated with a machine, themselves organized in clusters, and so on. Formally, a *hierarchy* is a set $\mathscr{H}(S) = \{S_1, \ldots, S_p\}$ of subsets of S that contains the whole resource set ($S \in \mathscr{H}(S)$), each singleton ($\forall s \in S, \{s\} \in \mathscr{H}(S)$), and such that any two parts in $\mathscr{H}(S)$ are either disjoint or included one in another ($\forall (S_i, S_j) \in \mathscr{H}(S)^2$, either $S_i \cap S_j = \emptyset$ or $S_i \subset S_j$ or $S_i \supset S_j$). A hierarchy is thus equivalent to a *rooted tree* where the nodes correspond to these hierarchical subsets, the *leaves* corresponds to the singletons, the *root* to the whole set, and the *tree-order* to the subset relation.
- The *event type* dimension is defined as the set $U = \{u_1, \ldots, u_l\}$ of the different possible types of events that occur during the trace execution.

We define our microscopic models as 3D matrices M^μ with dimensions $|S| \times |T| \times |U|$, and we name $v_{(s,t,u)}$ each entry associated with coordinates (s, t, u)

We propose different metrics that can be used to fill the microscopic model from the data contained in the raw trace:

- *Event occurrence number* corresponds to the number of events of a certain type u that occurs during a time interval t and generated by a resource s. This is the simplest approach.
- *State duration* corresponds to the total time passed in a particular state type u (a state is an event, like a function call, associated with start and end timestamps) over a time interval t, for a given resource s.
- *Variable mean* is the mean value of a variable associated with a resource s over a time interval t, and of a given type u.

3.4 Temporal Overview

In order to detect temporal anomalies in the application behaviors, our first contribution was a temporal overview for trace analysis fitting with Lamarche-Perrin methodology.

3.4.1 Methodology Application

1. We aggregate over time dimension.
2. The operands are the time periods $t \in T$. For each time period $t \in T$, we associate a value constituted by the sub matrix $M^{\mu}(t)$, with the dimensions $|S| \times |U|$.
3. We constrain the aggregation by allowing only the aggregation of contiguous time periods ($\mathscr{I}(T)$), without overlapping.
4. The result of time period aggregation is the sum of their associated matrices: $M^{\mu}(T_{(i,j)}) = \sum_{t \in T_{(i,j)}} M^{\mu}(t)$.

The algorithms and complexity measure calculation are presented in our previous works [3, 4, 13]. The aggregation steps are the following: the algorithm first calculates the quality measures of each possible time interval. Then, the user provides a value for p. The aggregation algorithm gives in return the best partition associated with p. To do that, we use dynamic programming and take advantage of the partition lattice structure and the additivity of their gain, loss and pIC, which helps us to decompose this set partition problem.

3.4.2 Visualization

The algorithm output is the set of aggregated time periods as a function of the parameter p. In order to provide a visualization that brings meaning to the user, we show a stacked bar chart, where the time is used as abscissa. Each stack of rectangles is associated with a temporal aggregate, organized following the same order over time. Its width is determined by the number of time slice aggregated. Each layer is associated with a type u. Its amplitude is computed as followed:

$$Q(T_{(i,j)}, u) = \frac{\sum_{t \in T_{(i,j)}} \sum_{s \in S} v_{(s,t,u)}}{j - i + 1} \tag{4}$$

In the case where layers are too small to be printed correctly, e.g., their sizes are less than one pixel, we visually aggregate them and show them differently.

3.4.3 Examples

In our previous works [3, 4, 13], we describe examples of applications where the temporal overview shows anomalies. For instance, we perturbed a GStreamer multimedia application playing a video with a CPU stress. The perturbation is easily detectable with Ocelotl (Fig. 2). We also analyzed MPI applications where we expected a regular temporal behavior. We were able to detect some perturbations provoked by a concurrency access to the network with other user running applications on the same platform (Figs. 3 and 4).

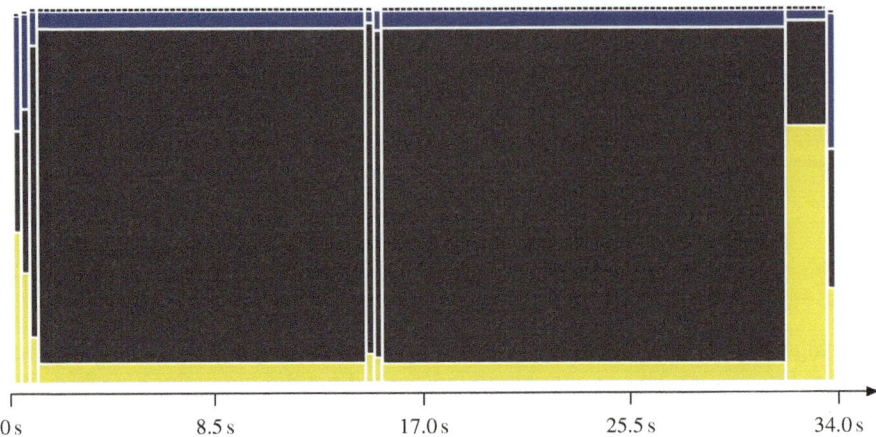

Fig. 2 Temporal overview showing a GStreamer multimedia application perturbed by a CPU stress (around 15 s). State colors correspond to the following GStreamer debug levels: *Grey* – LOG, *Blue* – DEBUG, *Yellow* – WARNING

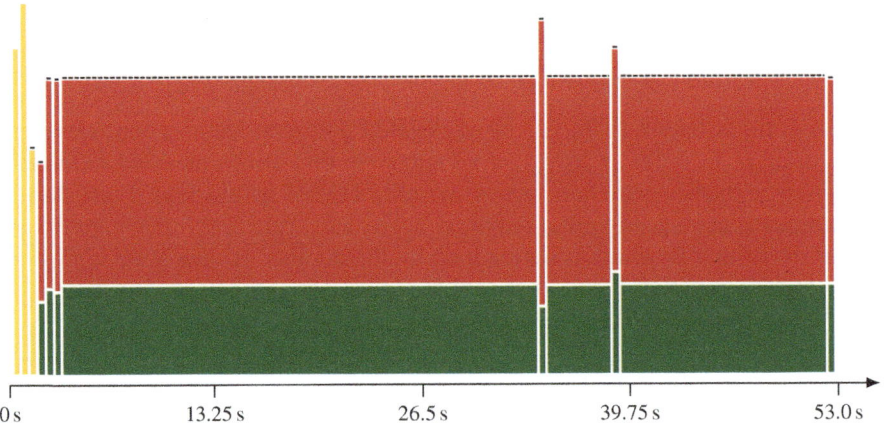

Fig. 3 Temporal overview showing a MPI application (NASPB CG class C, 64 processes) with two perturbations (around 35 and 40 s) provoked by a concurrency access to the network. State colors correspond to following the MPI calls: *Yellow* – MPI_Init, *Green* – MPI_Send, *Red* – MPI_Wait

3.5 Spatiotemporal Overview

Temporal overview can show temporal anomalies, but is limited to represent phenomena that also involve the application structure. This issue is at the bottom of the design of a new overview, showing both space and time.

3.5.1 Methodology Application

1. We aggregate **simultaneously** over time and space dimension.
2. The operands are the area determined by the Cartesian product of the time periods $t \in T$, and space elements $s \in S$. For each couple (s, t), the associated metric is the row matrix $M^\mu(s, t)$ with the dimension $|U|$.
3. We constrain the aggregation as the Cartesian product of the contiguous time periods $\mathscr{I}(T)$ and the hierarchy $\mathscr{H}(S)$, without overlapping.
4. The result of spatiotemporal elements aggregation is the sum of the associated vectors: $M^\mu(S_k, T_{(i,j)}) = \sum_{s \in S_k} \sum_{t \in T_{(i,j)}} M^\mu(s, t)$.

The corresponding algorithm and complexity measure calculation are presented in our previous works [2]. Similarly to what we did for temporal aggregation, we first compute the quality measures, but this time, for each time interval associated with an element of the hierarchy (leaves or nodes). Then, the user provides a value for p and the algorithm returns the best associated partition. The algorithm takes advantage of the same properties than for the temporal aggregation (dynamic programming, lattice structure, quality measure additivity). It is mainly constituted by iterative-recursive sequences searching successively for spatial partitions and temporal cuts.

3.5.2 Visualization

We represent as algorithm output the spatiotemporal area that are aggregated, as a function of a parameter p. With each area, we associate a color, that corresponds to the event type $u_{mode} \in U$ such as:

$$M^\mu(S_k, T_{(i,j)}, u_{mode}) = \max_{u \in U} \left(\frac{\sum_{t \in T_{(i,j)}} \sum_{s \in S_k} v_{(s,t,u)}}{|T_{(i,j)}| \times |S_k|} \right) \tag{5}$$

When the number of resources $|S|$ is greater than the amount of pixels in the spatial axis, we use visual aggregation (during rendering) to keep a clean visualization: if an aggregate has a visual height inferior to a threshold (in pixels), its parent in the spatial hierarchy is drawn instead. Visually-aggregated areas are marked differently: by a diagonal line, if underlying resources have the same temporal data partitioning, or by a cross otherwise. Figure 5 shows these different aggregation patterns.

3.5.3 Example

We mainly provide MPI application use cases to show the relevancy of our spatio-temporal overview technique. In these cases, we still show temporal perturbations

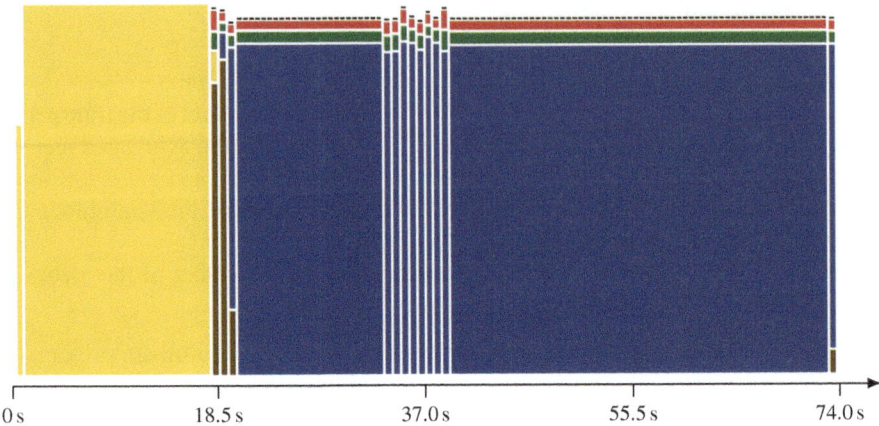

Fig. 4 Temporal overview showing a MPI application (NASPB LU class C, 700 processes) with one perturbation (37 s) provoked by a concurrency access to the network. State colors correspond to the following MPI calls: *Yellow* – `MPI_Init`, *Brown* – `MPI_Allreduce`, *Green* – `MPI_Send`, *Red* – `MPI_Wait`, *Blue* – `MPI_Recv`

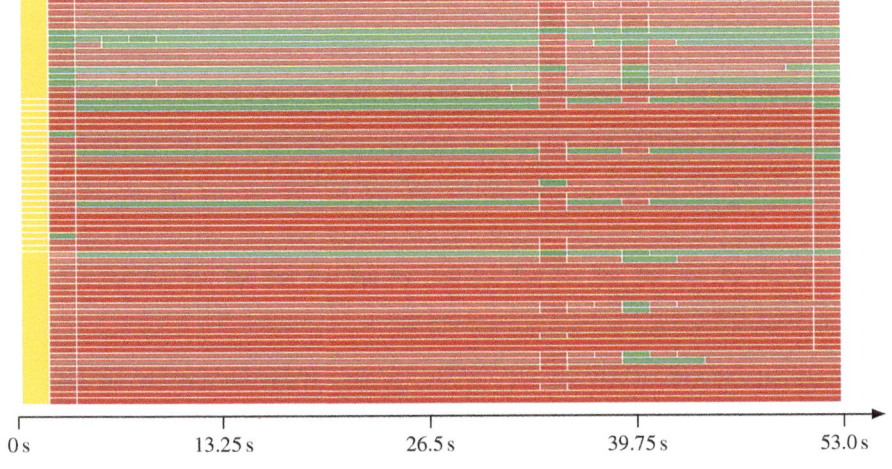

Fig. 5 Spatiotemporal overview of the application shown by Fig. 3. We still see the two temporal perturbations (around 35 and 40 s). State colors correspond to following the MPI calls: *Yellow* – `MPI_Init`, *Green* – `MPI_Send`, *Red* – `MPI_Wait`

that can be detected with the temporal algorithm (Fig. 4), but also spatial hetero-geneity that indicates that resources behave differently, which can be a problem (Fig. 5).

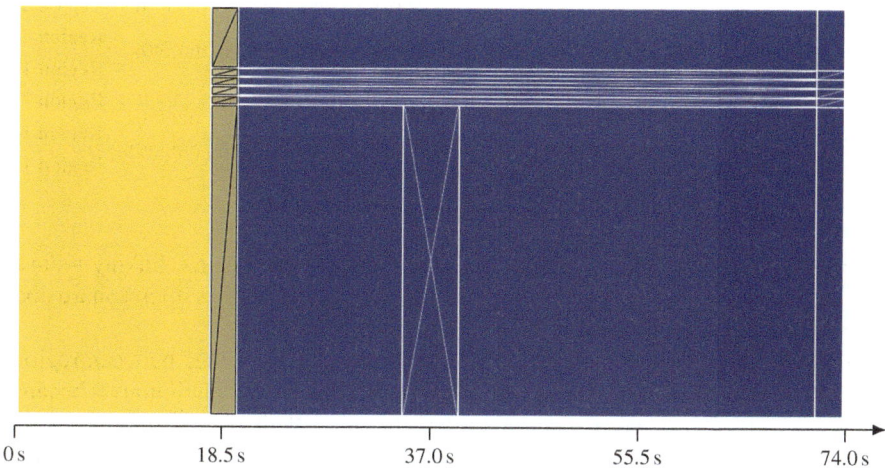

Fig. 6 Spatiotemporal overview of the application shown by Fig. 6. We find the temporal perturbation again (37 s) but now, we can see that it concerns only one cluster. Moreover, another cluster behaves heterogeneously during the computation phase. State colors correspond to the following MPI calls: *Yellow* – MPI_Init, *Brown* – MPI_Allreduce, *Blue* – MPI_Recv

4 Ocelotl Tool

In this section, we present Ocelotl, an analysis tool that implements the overviews presented in Sect. 3. We first describe its global architecture. In particular, we give details about the interaction features that provide the capability of a dynamic analysis. We then insist on the interactivity and the performance by showing techniques we use to reduce the computation times.

4.1 Global Architecture

Ocelotl is written in Java and is an Eclipse plug-in. It is a module of Framesoc [3], a trace and tool manager and analysis infrastructure. Framesoc provides an API to access traces, stored as data bases using a generic data model. Each trace is imported only once into Framesoc, and remains accessible for the following analysis sessions. Framesoc also provides a graphical interface and a bunch of tools (statistics, Gantt chart-like space-time diagram) classically used in the trace analysis domain.

4.1.1 Component Organization

Ocelotl is composed of a core and a graphical interface. The part related to the aggregation algorithm is a shared library we have written in C++ and interfaced through

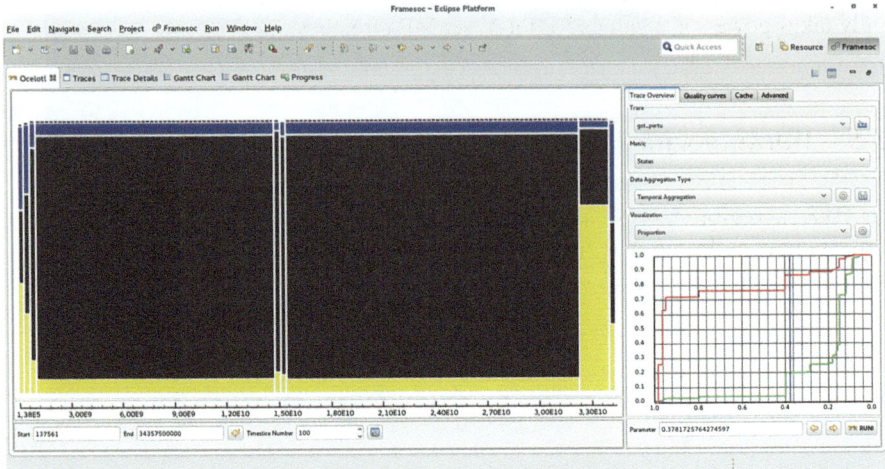

Fig. 7 Screenshot of Ocelotl showing its graphical interface. On the *left*, the overview composed by a temporally aggregated visualization. On the *bottom right*, curves show the complexity of the different possible representations (*in green*) and the information quantity contained inside (*in red*)

JNI (*Java Native Interface*). The libraries are named lpaggreg and lpaggregjni. This design choice was motivated by performance reasons (computation time, accurate memory management).

The core is responsible of the computations. The graphical interface provides various settings to the user. An area is dedicated to the rendering of the algorithm output. Another one is used to represent gain and loss curves that help to browse through the different aggregations. A screenshot of the GUI is shown in Fig. 7.

Ocelotl is relatively modulable; we provide three types of extension points:

- *microscopic model*: an extension that generates a microscopic model as a matrix over time, space and type dimensions. The currently implemented metrics to fill the matrices are the *event occurrence number*, the *state duration*, or the *variable mean* (cf. Sect. 3.3). Other metrics can be added by providing new extensions using the interface we provide.
- *aggregation type*: an extension that determines on which dimension aggregate and which algorithm of the C++ library is chosen. We currently propose *temporal* and *spatiotemporal* aggregations. As for microscopic models, new extensions can be added.
- *visualization*: an extension to draw the algorithm output. For the temporal aggregation, we draw the output as a simple partition using colored rectangles or as a stacked bar chart (cf. Sect. 3.4.1). For the spatiotemporal aggregation, we represent the aggregated area, or the mode (cf. Sect. 3.5.1). Other visualizations can also be added.

These extensions are dynamically detected when Ocelotl is launched. The user chooses, thanks to combo boxes, which microscopic model, which aggregation type, and which visualization he wants.

4.1.2 Pipelined Analysis Process

The analysis process is a pipeline composed of different steps:

1. The analyst tunes the different settings (choice of the trace, microscopic model, aggregation type, visualization, time bounds, resources and event types, time slice number and so on.) in the graphical interface. Extensions can also have their own settings. Once done, the user launches the analysis.
2. Ocelotl generates a query using the parameters defined by the user, provides it to Framesoc and gets a result set containing the events matching with the query.
3. Ocelotl fills the microscopic model, whose dimensions are determined by step 1 settings, while reading the result set. This process is multithreaded.
4. Ocelotl provides the microscopic model to the aggregation algorithm library.
5. Ocelotl calls the quality computation function of the shared library. This function computes the gain and loss associated with each possible aggregates.
6. Ocelotl calls the bissection function of the library, which returns a list of parameter p values, such as each partition P of this list is unique. This list is used to build an interactive graph of the gain and loss associated with P as a function of p. This graph is shown on the bottom right of the Fig. 7.
7. The user can click on this graph to select a value of p.
8. Ocelotl calls the algorithm function that provides the partition for a value of p in input. Since the user can manually provide a parameter p that is not contained in the list, this partition is recomputed and not retrieved from the step 6.
9. Ocelotl generates a visualization from the algorithm output and the microscopic data.
10. The user can change the partition by choosing another p value. Back to the step 8.
11. The user can change other settings (change time bounds to zoom, modify resources involved, change time slice number to generate microscopic model, etc.). Back to the step 1.

4.1.3 Interaction

We provide different types of interaction in Ocelotl. First, as presented in Sect. 4.1.2 (step 7), the user can click on the gain and loss curves in order to select a parameter p. The objective is to choose the most interesting values of p thanks to the curve shapes. Second, we provide a zoom mechanism in the visualization. By selecting an area with the mouse, the user changes the time bounds and can relaunch an analysis on this area. The pipelined process is started from the beginning: indeed, the zoom

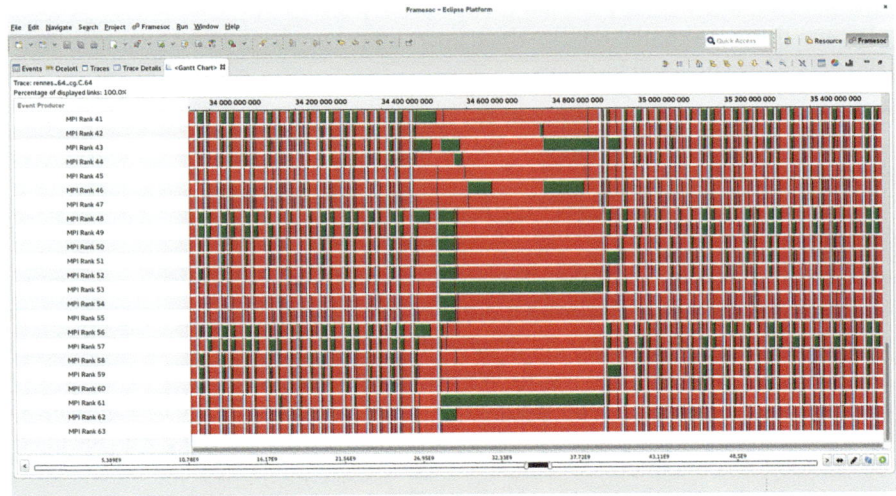

Fig. 8 Example of the Gantt chart provided by Framesoc, focused on one of the temporal perturbations detected by Ocelotl in the use case NASPB CG class C, 64 processes (Fig. 3)

is not graphical, but performs a reiteration of the full process on the selected subpart of the trace. The new visualization obtained this way is thus more accurate. Last, the user can switch to a space-time Gantt chart-like diagram. By clicking on a button, Framesoc opens a new Gantt chart window on the selected time bounds. Here, the objective is to provide more details on an interesting time part detected thanks to Ocelotl. An example of Gantt chart is given by Fig. 8, showing a perturbation detected by Ocelotl.

4.2 Performance

Regarding the Ocelotl performances, three points have to be evaluated separately:

- The microscopic model building: it corresponds to the steps 1–4 of the pipelined process. It consists in reading the trace and getting its events through a query to the database, and filling the microscopic model using aggregation. Figure 9 shows that for a given time slice number, the time of microscopic model building from the database is linear to the event number. As trace reading is a costly process, we have implemented a way to save microscopic models associated with a trace, which reduces dramatically the microscopic model building time for further analysis sessions (reading the trace is mandatory at least once). We describe this feature in Sect. 4.2.1.

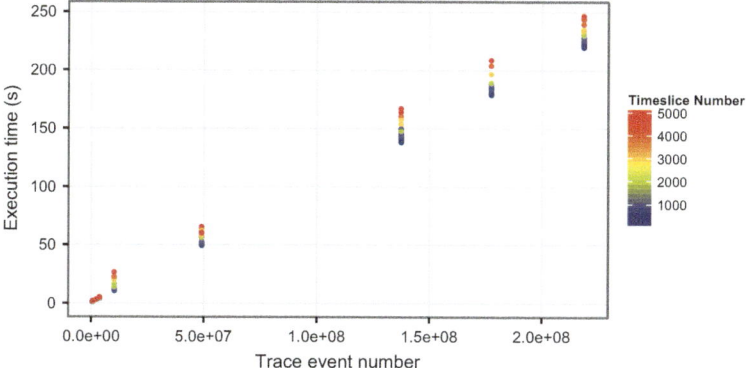

Fig. 9 Microscopic building computation time, including the trace reading from the database, as a function of event number contained in the trace. Time slice number is symbolized by a color gradient. For a given time slice number, execution time is linear to the event number. Most part of execution time is due to database trace reading

- The algorithm computation: it depends mainly on the microscopic model dimensions. In the temporal algorithm case, the costliest part is the quality computation, whose complexity is $\mathcal{O}(|T|^2 \times |S| \times |U|)$. The time slice number is the main parameter the user tunes. Increasing its value provides more precision, but leads to longer computation times. In the spatiotemporal algorithm case, quality computation and bissection time have the same order of magnitude. Its total complexity (sum of the different operations) is $\mathcal{O}(|T|^3 \times |\mathcal{H}(S)|)$. In practical, $|T|$ should be left less than 30 time slices to keep reasonable computation delays. In both temporal and spatiotemporal cases, part computation times are marginal compared to quality and bissection times. Figures 10 and 11 show the algorithms execution time as a function of the microscopic model size. Bissections are computed with a maximum difference of 0.001 between two values of p.
- The rendering: the time needed to build the visualization from the algorithm output. It depends mainly on the graphical object number, and is thus related to the time slice number in the case of the temporal overview, and to the space and time dimension elements in the case of the spatiotemporal overview. In the worst cases, this time reaches about 10 s. Responsiveness depends partially on the SWT and Draw2D Eclipse libraries performances.

Table 1 shows Ocelotl execution times for real MPI traces of different sizes. All our tests have been performed using a desktop computer, composed of 4 cores Intel Xeon CPU E3-1225 v3 at 3.20 GHz, 32 GB DDR3, 256 GB SSD. All the traces and their caches were stored on the SSD, which had a SATA-3 interface and a reading capacity of 530 MB/s.

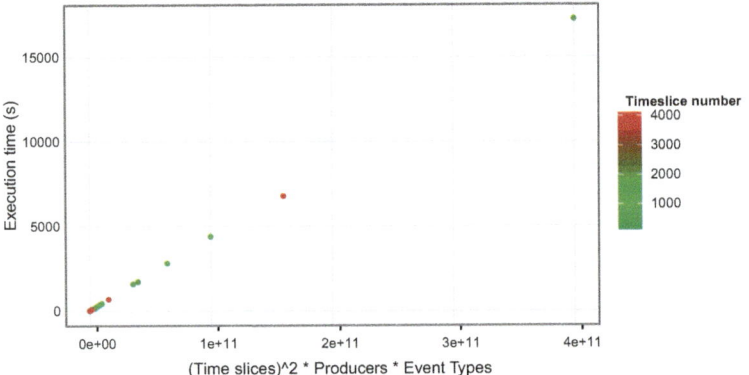

Fig. 10 Total computation time of the temporal aggregation algorithm. We represent execution time as a function of $|T|^2 \times |S| \times |U|$ to highlight the algorithm complexity. As an indication, the complexity of the temporal analysis of NASPB LU class C, 700 processes (Fig. 6) is $|T|^2 \times |S| \times |U| = 100^2 \times 700 \times 10 \sim 10^8$

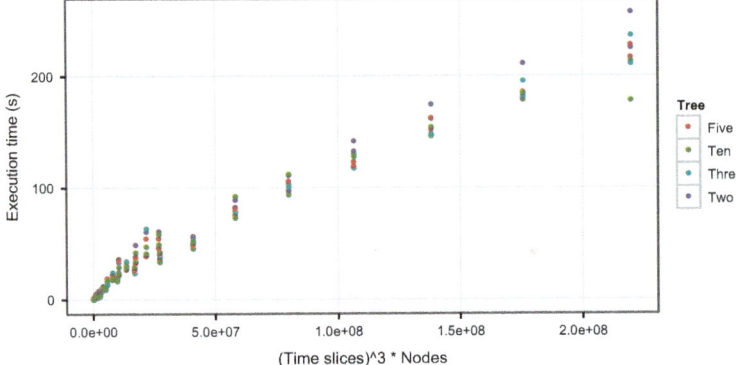

Fig. 11 Total computation time of the spatiotemporal aggregation algorithm. We represent execution time as a function of $|T|^3 \times |\mathscr{H}(S)|$ to highlight the algorithm complexity. Tree corresponds to the hierarchy structure (number of child per node). As an indication, the complexity of the spatial analysis of NASPB LU class C, 700 processes (Fig. 5) is $|T|^2 \times |\mathscr{H}(S)| = 30^3 \times 801 \sim 2 \times 10^7$

4.2.1 Trace Reading Time Reduction

Reuse Generated Microscopic Models

In order to speed up the access to a trace data, we implemented a cache. This cache works by saving in a file the microscopic model the first time it is built. Since all the computations performed by Ocelotl are based on this microscopic model, the cache avoids the necessity to go through the costly process of reading the whole trace in the database in later analysis sessions. The microscopic model is based on an aggregation of the events which depends on the used metric (i.e. event occurrence,

Table 1 Ocelotl execution times for various trace sizes (temporal analysis, $|T| = 100$)

	Case A	Case B	Case C	Case D
Application	CG, class C	CG, class C	LU, class C	LU, class B
Processes	64	512	700	900
Site	Rennes	Grenoble	Nancy	Rennes
Event number	3,838,144	49,149,440	**218,457,456**	177,376,729
Trace size (in DB)	136.9 MB	1.8 GB	**8.3 GB**	6.7 GB
Ocelotl computation times				
Trace reading + Microscopic model	4 s	50 s	**224 s**	181 s
Qualities computation	<1 s	1 s	**2 s**	3 s
Bissection	<1 s	<1 s	**<1 s**	<1 s
Total preprocess	4 s	51 s	**226 s**	184 s
Parts computation + Rendering	<1 s	<1 s	**<1 s**	<1 s

state duration, variable mean) and, for a given metric, this microscopic model does not change from one analysis to another, thus allowing its reuse. Thanks to the aggregation, the size (memory footprint) of a microscopic model is much smaller than the raw trace stored in database and consequently is much faster to load. To further reduce the footprint and improve performance, we only save non-null values of the microscopic model. As an illustration of the obtained footprint reduction, a trace with 1 billion events takes 35.6 GB in database but takes only 34.8 MB as a cache file of 1,000 time slices.

As explained in Sect. 3.3, the size of a microscopic model depends on three dimensions: the number of time slices, the number of event producers and the number of event types (i.e. the temporal, the spatial and event type dimensions respectively in Sect. 3.3). However the number of time slices is the only dimension that has an influence on the values stored in the microscopic model. Indeed, modifications in the number of event producers or in the number of event types can be dealt with in the later computations by simply ignoring the stored values corresponding to the ignored event producers or event types. As a consequence, the saved microscopic model is built with a high number of time slices,[1] in order to have a microscopic model with a high granularity. This allows us to use the cache for all the analyses performed with a number of time slices lower than the one used in the cached microscopic model. For example, suppose that a cache was generated with 1,000 time slices ($|T| = 1,000$), and the analyst wants to perform an analysis with 100 time slices ($|T| = 100$). Then a time slice of the microscopic model will be build by using 10 time slices of the cached one. The operation performed for rebuilding is a simple addition of the values of the cached time slices. More

---

[1] This number is configurable through the GUI in Ocelotl.

a *Example of microscopic model rebuilding using perfect cache*

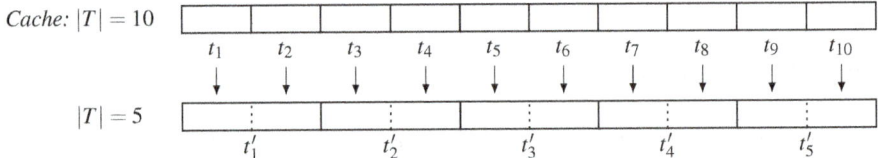

b *Example of microscopic model rebuilding with dirty time slices (in red)*

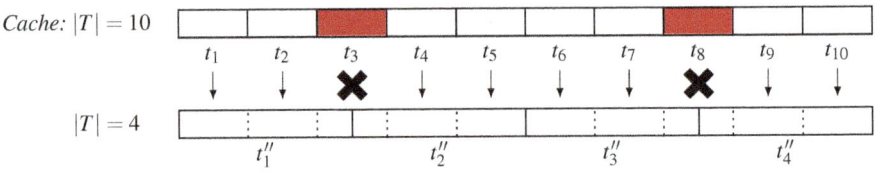

Fig. 12 Illustration of the rebuilding of a microscopic model using a 10 time slice cached microscopic model: (**a**) Cache time slice number fits with generated microscopic model (**b**) Cache time slice number does not fit, which is responsible of dirty time slices

formally, for a time slice t_{built} using the cached time slices from t_{cache_i} to t_{cache_j}, the value of t_{built} is:

$$t_{built} = \sum_{n=i}^{j} t_{cache_n} \tag{6}$$

This is illustrated by the Fig. 12a: in this example, each new time slice is composed of two time slices of the cache, e.g. $t_1' = t_1 + t_2$. Figure 13 shows the improvement when using a presaved microscopic model instead of reading the trace events.

Generate an Approximated Microscopic Model

Rebuilding the microscopic model works well when the number of time slices in the cached microscopic model is divisible by the number of time slices used in the newly built microscopic model. However, when this is not the case, it can lead to problematic cases where some of the cached time slices are split over two time slices. This phenomenon is illustrated in the Fig. 12b. In this figure, we can see that the time slices t_3 and t_8 each overlaps over two time slices, and thus cannot be used "as is" to build the new microscopic model. We call these cache time slices split across two time slices, "dirty" time slices. This problem can also occur when performing a zoom on a region of the trace. To deal with those problems, we implement an approximate strategy. The strategy is a tradeoff between speed and

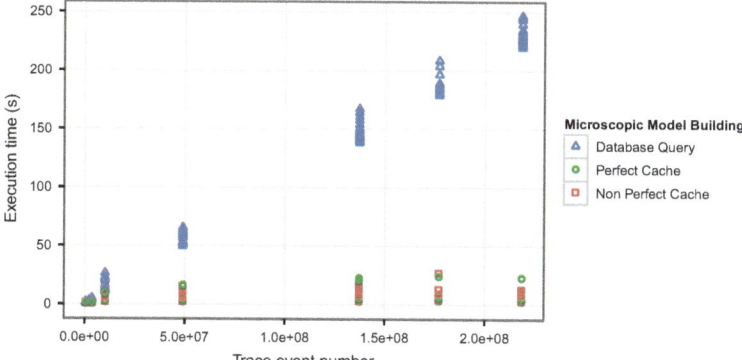

Fig. 13 Microscopic building computation time as a function of event number contained in the trace. Time slice number varies from 10 to 200. We distinguish several cases: events are retrieved by a *query* to the database to build the microscopic model, or microscopic model is generated from a previously saved data cache. We distinguish the case where both microscopic model time slices fit perfectly (*perfect*), and the case where an approximation has to be done (*non perfect*)

accuracy. It consists in computing for each dirty time slice, the proportions of the time slice which overlaps the currently built time slices, which gives a value between 0 and 1. We then multiply the values of the dirty time slice, with the corresponding proportion. More formally, with the notations introduced in Sect. 3.3, for a given dirty time slice t_{dirty} and the two built time slices t_{built1} and t_{built2} it overlaps, we compute $prop_1$ and $prop_2$, with:

$$prop_1 = \frac{\tau_{end}(t_{built1}) - \tau_{start}(t_{dirty})}{d(t_{dirty})} \qquad (7)$$

where $\tau_{start}(t)$ and $\tau_{end}(t)$ correspond respectively to the start and the end timestamp of the timeslice t. We can then compute $prop_2$ as $prop_2 = 1 - prop_1$. Then for each value in the time slice t_{built1} in the built microscopic model, we perform the following operation:

$$v_{built(s,t_{built1},u)} = v_{built(s,t_{built1},u)} + v_{cache(s,t_{dirty},u)} \times prop_1 \qquad (8)$$

and a similar operation for t_{built2} with $prop_2$.

This strategy provides good results in term of performance (Fig. 13), and only has a small overhead compared to using a cache with no dirty time slices. However, the performed approximations can lead to discrepancies between the microscopic model obtained and the original one. In our experiments, we have observed that, when existing, the large majority of the differences are minor, as illustrated by Fig. 14, and do not hinder the analysis results. However in some rare cases, especially when

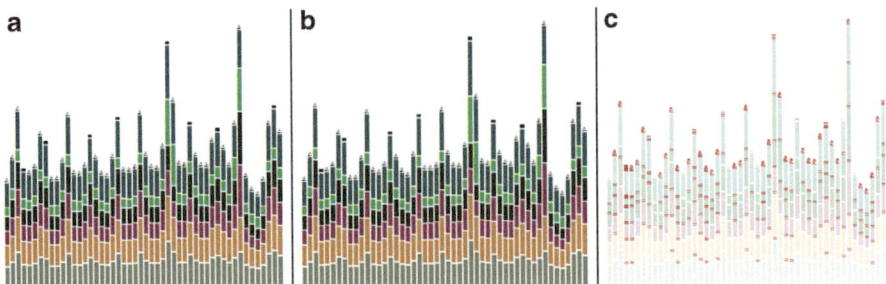

Fig. 14 Example of the differences obtained with the approximate strategy: (**a**) is the view obtained when using no cache, (**b**) is the view obtained when using the approximate strategy and (**c**) highlights in *red* the differences between the two views

the ratio of dirty time slice number to time slice number is high, more problematic cases can appear, leading to different aggregations. This is why this strategy should be used with caution and used to quickly explore the trace. In the case where the user wants to keep maximum fidelity in the microscopic model, an option can be selected to get the events from the database trace when the cache does not fit well, instead of using the approximate strategy.

5   Conclusion

We have presented Ocelotl, a visualization tool that provides multidimensional overviews for trace analysis. These overviews are built upon an aggregation methodology that involves information quantities, and gather trace parts that behave homogeneously. This method enables to reduce the visualization complexity while keeping useful information and highlighting application phases and anomalies. We present various examples showing the relevancy of our temporal and spatiotemporal overviews. We succeed in finding anomalies in 8 GB traces containing up to 218 millions events.

The Ocelotl tool has features that help the analysis. Its modularity allows to add new components adapted to particular needs. Interaction mechanisms like zoom and switches to other representations help to get easily more details on subparts of the trace. Regarding the performance and the interactivity, we profit from the Framesoc framework that stores traces in databases. We succeed in reading 1 billion event traces of 35 GB in 18 min. As reading the trace remains globally a costly process, we save intermediate trace abstractions, the *microscopic models*, for further analysis sessions. This enables to get a result in up to 30 s, whatever the original trace size and event number.

Our future work concerns many domains. Aggregation methodology could be extended on other dimensions to represent trace aspects like communications, which we are unable to manage for now. Some efforts have to be done on the visualization, in particular by providing dynamically more information about the aggregates. Regarding the performance, the aggregation algorithms could be parallelized, which could help us to be more accurate, in particular on the temporal dimension of spatiotemporal overview.

References

1. Chassin de Kergommeaux, J.: Pajé, an interactive visualization tool for tuning multi-threaded parallel applications. Parallel Comput. **26**(10), 1253–1274 (2000)
2. Dosimont, D., Lamarche-Perrin, R., Schnorr, L.M., Huard, G., Vincent, J.M.: A spatiotemporal data aggregation technique for performance analysis of large-scale execution traces. In: Proceedings of the 2014 IEEE International Conference on Cluster Computing (CLUSTER'14), Madrid (2014)
3. Dosimont, D., Pagano, G., Huard, G., Marangozova-Martin, V., Vincent, J.M.: Efficient analysis methodology for huge application traces. In: Proceedings of the 2014 International Conference on High Performance Computing & Simulation (HPCS), Bologna (2014)
4. Dosimont, D., Schnorr, L.M., Huard, G., Vincent, J.M.: A trace macroscopic description based on time aggregation. Technical report HAL RR-8524 (2014)
5. Knüpfer, A., Brunst, H., Doleschal, J., Jurenz, M., Lieber, M., Mickler, H., Müller, M.S., Nagel, W.E.: The Vampir performance analysis tool-set. In: Tools for High Performance Computing, pp. 139–155. Springer, Berlin (2012)
6. Kullback, S., Leibler, R.: On information and sufficiency. Ann. Math. Stat. **22**(1), 79–86 (1951)
7. Lamarche-Perrin, R., Demazeau, Y., Vincent, J.M.: How to build the best macroscopic description of your multi-agent system? In: Demazeau, Y., Ishida, T. (eds.) Proceedings of the 11th International Conference on Practical Applications of Agents and Multi-Agent Systems (PAAMS'13), Salamanca. LNCS/LNAI, vol. 7879, pp. 157–169. Springer, Berlin/Heidelberg (2013)
8. Lamarche-Perrin, R., Demazeau, Y., Vincent, J.M.: The best-partitions problem: how to build meaningful aggregations. In: Proceedings of the 2013 IEEE/WIC/ACM International Conference on Intelligent Agent Technology, Atlanta, pp. 399–404 (2013)
9. Lamarche-Perrin, R., Demazeau, Y., Vincent, J.M.: Building the best macroscopic representations of complex multi-agent systems. In: Transactions on Computational Collective Intelligence. LNCS. Springer, Berlin/Heidelberg (2014)
10. Lamarche-Perrin, R., Schnorr, L.M., Vincent, J.M., Demazeau, Y.: Evaluating trace aggregation for performance visualization of large distributed systems. In: Proceedings of the 2014 IEEE International Symposium on Performance Analysis of Systems and Software, Monterey, pp. 139–140 (2014)
11. Linux Tools Project/LTTng2/User Guide – Eclipsepedia. http://wiki.eclipse.org/index.php/Linux_Tools_Project/LTTng2/User_Guide (2015)
12. Lusk, E., Chan, A.: Early experiments with the OpenMP/MPI hybrid programming model. In: OpenMP in a New Era of Parallelism. LNCS, vol. 5004, pp. 36–47. Springer, Berlin (2008)
13. Pagano, G., Dosimont, D., Huard, G., Marangozova-Martin, V., Vincent, J.M.: Trace management and analysis for embedded systems. In: 2013 IEEE 7th International Symposium on Embedded Multicore Socs (MCSoC), Tokyo, pp. 119–122 (2013). doi:10.1109/MCSoC.2013.28

14. Pillet, V., Labarta, J., Cortes, T., Girona, S.: Paraver: a tool to visualize and analyze parallel code. In: Proceedings of WoTUG: Transputer & Occam Developments, Manchester, vol. 44, pp. 17–31 (1995)
15. Prada-Rojas, C., Riss, F., Raynaud, X., De Paoli, S., Santana, M.: Observation tools for debugging and performance analysis of embedded linux applications. In: Conference on System Software, SoC and Silicon Debug-S4D, Sophia-Antipolis (2009)
16. Schnorr, L.M., Legrand, A., Vincent, J.M.: Detection and analysis of resource usage anomalies in large distributed systems through multi-scale visualization. Concurr. Comput. 24(15), 1792–1816 (2012)
17. Shannon, C.E.: A mathematical theory of communication. Bell Syst. Tech. J. 27(3), 379–423, 623–656 (1948)

Integrating Critical-Blame Analysis for Heterogeneous Applications into the Score-P Workflow

Felix Schmitt, Robert Dietrich, and Jonas Stolle

Abstract High performance computing (HPC) systems increasingly deploy accelerators and coprocessors to achieve maximum performance combined with high energy efficiency. Thus, application design for such large-scale heterogeneous clusters often requires to utilize multiple programming models that scale both within and across nodes and accelerators. To assist programmers in the complex task of application development and optimization, sophisticated performance analysis tools are necessary. It has been shown that CASITA, an analysis tool for complex MPI, OpenMP and CUDA applications, is able to effectively identify valuable optimization targets by means of critical-blame analysis for applications utilizing multiple programming models. This paper presents the integration of CASITA into the Score-P tool infrastructure. We depict the complete Score-P measurement and analysis workflow, including the performance data collection for the CUDA, OpenMP and MPI programming models, tracking of dependencies between work performed on the host and on the accelerator as well as waiting-time and critical-blame analysis with CASITA and visualization of analysis results in Vampir.

1 Introduction

Since the fastest supercomputer in the 2010 top 500 list [23] deployed GPUs as hardware accelerators, the number of hybrid system is increasing. Currently the systems with accelerators or coprocessors (12.8 %) yield 34.5 % of the theoretical peak performance. Due to the complexity of these systems their programming is challenging. Several programming interfaces (e.g. MPI, CUDA, OpenMP) have to be applied and supported by development and analysis tools. CASITA, a tool for critical-blame analysis for heterogeneous, distributed applications, has been developed to guide developers to critical optimization targets in their programs. The critical-blame analysis bases on two fundamental properties of parallel programs: the critical path and load-imbalances. It has been proven in [21] that these prop-

F. Schmitt • R. Dietrich (✉) • J. Stolle
Center for Information Services and High Performance Computing, Technische Universität
Dresden, 01062 Dresden, Germany
e-mail: felix.schmitt@tu-dresden.de; robert.dietrich@tu-dresden.de; jonas.stolle@tu-dresden.de

© Springer International Publishing Switzerland 2015   161
C. Niethammer et al. (eds.), *Tools for High Performance Computing 2014*,
DOI 10.1007/978-3-319-16012-2_8

erties are important to identify bottlenecks in complex heterogeneous applications. CASITA incorporates both into its analysis and final score that rates activity types for the potential to optimize the total program execution. It supports the most commonly used paradigms in HPC: MPI, OpenMP and CUDA. With CUDA and OpenMP 4.0 offloading the two dominant accelerator or coprocessor types are supported [7, 20].

As the concepts of critical-path analysis and blame distribution are not new, other tools co-exist with overlapping functionality as well as complementary features. To not worsen the fragmentation of the tools landscape we integrate our analysis in the scorep workflow. The integration into scorep avoids redundant effort in development and maintenance of the measurement system. Furthermore, this lowers the initial hurdle to use the analysis and improves the user experience for the overall concept of Score-P. Using Score-P we build on a common data format (OTF2) [9], which is also the basis for the well-known Scalasca analysis.

The remainder of this paper is organized as follows: Sect. 2 introduces the concept of critical-blame analysis and its implementation in the tool CASITA. Section 3 presents the integration of this tool into the Score-P and Vampir measurement and analysis workflow. In Sect. 4, we cover related work of other tools from the domain of critical-path and blame performance analysis. Finally, Sect. 5 summarizes the contributions of this paper and investigates future work.

2 Background: Critical-Blame Analysis

The critical-blame analysis is based on two important properties of a parallel program, its *critical path* and the *blame* assigned to activities for causing waiting time. Locating the critical path is important to analyze the performance of parallel programs as shown by Yang and Miller [24]. Since it determines the program's execution time, optimizing activities on this path can improve the total runtime. As a result, knowledge about the critical path helps performance analysts to identify most relevant optimization targets. Derived metrics can prove even more purposeful to identify inefficiencies with a high potential to reduce the program execution time. Boehme et al. [3] present a scalable performance analysis method that uses performance indicators based on the critical path for an MPI execution trace.

A requirement for locating the critical path is the identification of waiting time in the parallel program which appears when an activity is stalled by another concurrent event. This commonly occurs in barrier operations and (mostly blocking) synchronization or communication functions. A wait state for a certain programming model can be identified by applying a set of pattern matching rules. Figure 1 illustrates the rule for the CPU waiting on a CUDA kernel. This specific rule is e.g. triggered by the end event of the CUDA stream synchronization.

Once wait states have been located, the blame metric can be computed as a quantification of the delay caused by each activity in the program execution. Late activities forcing other program parts to wait are therefore attributed with this

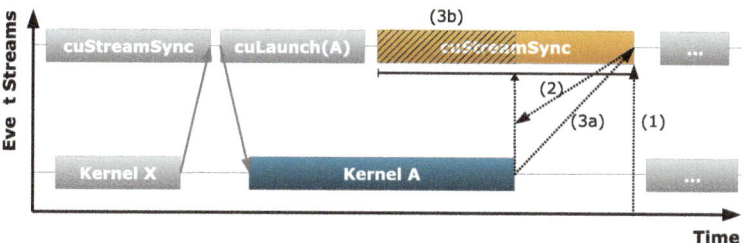

Fig. 1 CUDA Kernel Rule: On a blocking CUDA stream synchronization (1), the kernel stalling this activity is identified on the referenced CUDA stream (2). A dependency edge is inserted in the dependency graph (3a) and the waiting time is computed (3b) (Reprinted from [20])

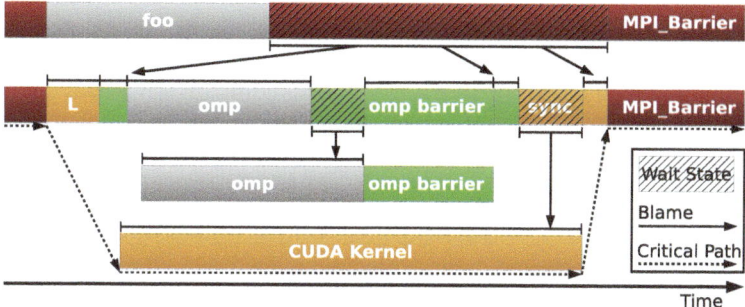

Fig. 2 Hierarchical Blame Distribution in CASITA: When distributing blame in traces containing multiple programming models (e.g. MPI, OpenMP and CUDA), it is important not to assign a penalty to activities for the time they are a wait state themselves (Reprinted from [21])

penalty. HPCToolkit applies the approach of blame shifting [22] to attribute blame to the potential root cause of a wait state. The performance analysis tool Scalasca [10] uses the term *cost of idleness* instead to identify performance-relevant activities in large-scale MPI applications.

The combination of critical-path and blame analysis has been investigated by Schmitt et al. in [20] for applications utilizing both the MPI and CUDA programming model. This concept has been further extended for non-nested OpenMP codes as well as compiler-instrumented regions. Its implementation in the critical-blame analysis tool CASITA has been presented in [21]. CASITA thereby respects the hierarchy of programming models. This avoids that blame for a wait state in one paradigm (e.g. MPI) is assigned to a wait state in another programming paradigm (e.g. OpenMP). Hierarchy-aware blame shifting is shown in Fig. 2. The tool computes the performance metric *critical blame* for an activity type (i.e. all instances of a particular function or kernel) from its exclusive time on the critical path combined with the aggregated blame, which is attributed to this activity for the overall waiting time it caused.

The metric critical blame quantifies regions on the critical path according to the load imbalance they induced in the parallel execution [21]. To compute wait states and distribute blame, CASITA uses programming-paradigm specific rules to insert dependency edges in its internal distributed event dependency graph. Each edge reflects the dependency introduced into the program flow by synchronization or communication operations such as MPI_Barrier. The analysis generates an optimization guidance report along with an OTF2 trace. This trace is enriched with counter information for waiting time, blame and the critical path and can be visualized in the trace browser Vampir [12].

3 Integrating CASITA into Score-P

This section depicts the analysis workflow we propose to perform critical-blame analysis. Starting with performance measurement using Score-P and requirements for the analysis with CASITA we describe the additional counter data generated by CASITA as well as their visualization in Vampir. We briefly present the new analysis capabilities with two use cases. Figure 3 shows where CASITA positions among the available analysis tools that base on the Score-P infrastructure.

3.1 Score-P Measurement Environment

Score-P [15] is a state-of-the-art performance measurement run-time infrastructure with a focus on HPC programming paradigms and applications. It evolved as a joined measurement backend for several previously separate tools, including Periscope, Scalasca, TAU and Vampir. With Score-P, these tools can make use of a unified data collection infrastructure, thereby significantly increasing the interoperability between them. The run-time measurement supports a large variety of data sources such as library wrapping, common tools interfaces (e.g. PMPI and CUPTI), external counters (e.g. PAPI) as well as compiler-aided instrumentation. Focusing on commonly used programming models in the domain of scientific HPC applications, Score-P can record data for MPI, OpenMP, CUDA, SHMEM and PThreads. Collected run-time information is stored in CUBE format profiles or OTF2 trace files. Both are open formats which can be used as input to multiple open source and proprietary analysis or visualization tools. Since the detection of an application's critical path requires its complete execution timeline, the remainder of this work is concerned with the creation of OTF2 trace files where all activities are captured and logged for each event stream.

Score-P consists of a frontend part, a set of external tools for the quick investigation of result files and a measurement backend. The latter comprises so called *adapters*, each handling data collection for a specific programming model. With respect to the CASITA critical-blame analysis tool, the most interesting adapters are for CUDA, OpenMP and MPI. Their implementation is covered in the following.

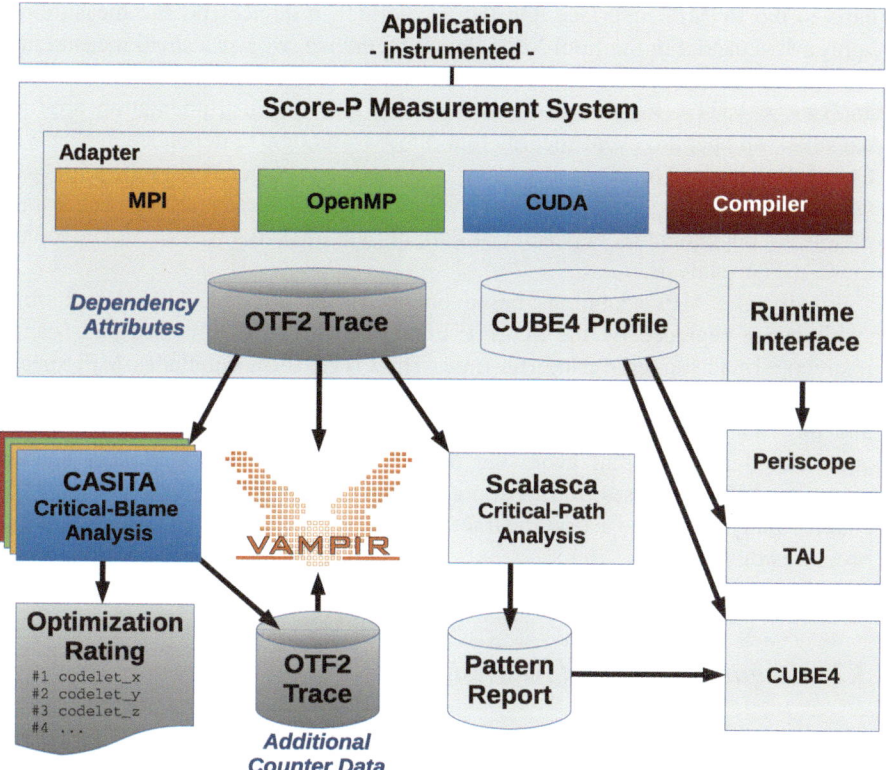

Fig. 3 Score-P Measurement and Analysis Workflow using CASITA: All commonly used paradigms in HPC are supported

The CUDA adapter's primary mean of data collection is the CUDA Profiling Tools Interface (CUPTI) [17], which provides several APIs aimed at performance tools to capture run-time information on CUDA programs. The *callback API* allows the tool to register for CUDA API calls on both the runtime and driver API level. Before and after the execution of a registered function in the CUDA driver a tool receives a callback including all function parameters. Score-P uses them to record performance events. Additionally, a tool is notified on resource-related events, e.g. the creation of a CUDA stream or context. In Score-P tracing mode CUDA streams are mapped to appropriate OTF2 locations in the trace file. Using CUPTI's *activity API* on the other hand, Score-P collects information on device-side activities, especially begin and end time stamps of kernels and asynchronous memory copies as well as their execution context. To handle different clocks on the host and the accelerator, Score-P integrates various time synchronization mechanisms.

For OpenMP 4.0, the adapter uses different data sources on the host and on the accelerator. On the host, the application source code is instrumented using OPARI2 [16]. It parses the code for OpenMP directives and inserts appropriate calls to the POMP2 interface. On the accelerator, a device-specific measurement approach is used. For the Intel Xeon Phi target device, we use a small measurement library [6] based on the OpenMP Tools Interface (OMPT) [8]. For ctitical-blame analysis, CASITA itself is not restricted to OPARI's instrumentation approach but relies on specific event records that allow to recognize dependencies between OpenMP threads. Those records carry information on OpenMP's fork-join execution model, allowing tools to identify which threads are spawned by each parallel region. As dependencies between tasks are not yet tracked, tasks are not covered in the CASITA analysis.

Finally, the MPI adapter is based on the PMPI interface [14]. Weak object symbols are replaced by the Score-P library so that MPI function calls can be intercepted and logged during run-time. The OTF2 format includes MPI-specific records that include, among others, information on the current MPI communicator and the root rank of a collective operation. OTF2 trace files consist of global and local definitions that avoid the need for unification of identifiers between recording MPI processes. Furthermore, its separation into multiple (logical) files enables analysis tools to access the trace concurrently, i.e. by using multiple analysis processes themselves.

3.2 Requirements and Analysis

CASITA requires dependency information embedded in the trace to reconstruct all runtime dependencies for the analysis. Such dependencies include for example the ranks participating in an MPI point-to-point or collective operation and the respective rank, the threads executing the same OpenMP barrier directive or the CUDA kernel initiated by a specific CUDA launch function call. For both MPI and OpenMP, all dependency information is already available in traditional OTF2 traces. In the case of MPI, OTF2 already provides specific records carrying model-specific information such as the participating MPI ranks. OpenMP is integrated similarly using records that capture its fork-join semantics. For CUDA, additional dependency attributes must be recorded at runtime to enable a subsequent critical-path and blame analysis. From a user's perspective, it is only required to set Score-P's environment variable SCOREP_CUDA_ENABLE to at least kernel,memcpy,driver,references before running the measured application. This notifies Score-P to collect records for CUDA kernels and memory copies using the CUPTI activity API and driver calls using the callback API.

Specifying references adds *OTF2 attributes* to certain CUDA records. There are currently three types of Score-P-specific attributes (called *keys*) that are required for a complete subsequent dependency analysis: *Stream keys* annotate which CUDA stream is referenced by a kernel launch, host-device or intra-device synchronization function. *Event keys* allow the tool to uniquely identify CUDA events issued to any stream for the purpose of synchronization, e.g. using cuEventSynchronize. *Result keys* capture the function return code of certain CUDA API calls such as `cuEventQuery` to store whether or not the call was successful. Those dependencies are available without additional performance measurement overhead as they are provided by the CUPTI callback API. However, some effort is required post-mortem for pattern matching in CASITA's analysis phase. Attributes are transparent to all other analysis tools that work with the respective trace file, e.g. Scalasca. The overhead for storing additional references in memory and on disk primarily depends on the rate of annotated CUDA API calls. It is typically very low but can be increasingly significant for frequently polling applications, e.g. an excessive use of `cuEventQuery`, for which the rate of API calls is very high, compared to the program execution time.

An OTF2 trace including additional CUDA dependency information can be used as input for CASITA's critical-blame analysis. It implements several phases. For all phases, the tool uses the same number of MPI analysis processes as in the original application, thus being a parallel application in its own right. Initially, the input trace is read from disk in parallel. Internal object representations for relevant events such as MPI or CUDA function calls are created. Second, a rule-based pattern matching approach is used to identify inefficiency patterns between multiple event streams. Dependency edges are added in the created *event dependency graph*. In the third phase, CASITA applies a parallel reverse replay to identify the critical path over MPI events. Re-enacting the original MPI communication structure has two advantages: It avoids the need for a global, unified dependency graph in favor of distributed sub-graphs that are only connected by the semantics of the MPI API. Moreover, the analysis exhibits similar scalability as the original program execution and can utilize the same number of HPC resources. Once global MPI critical path and blame have been computed, the same pattern matching is applied to critical sub-paths between two MPI events on the global critical path. Matching in all sub-paths located on one analysis process is parallelized using OpenMP. This creates the advantage that available compute resources are utilized more efficiently, even if the critical path is not distributed evenly over analysis processes. Finally, in the last phase, critical path and blame statistics are summarized to a *optimization guidance report*. For each activity type, i.e. all instances of a specific function call, the *critical blame* [21] metric is computed and used to rank activity types according to their potential for optimizing the parallel application. For comparison, exclusive runtime, critical-path time and blame on a percentage basis are printed for each activity type.

3.3   Visualization

Besides the report profile, the user can choose to re-write the original OTF2 trace with additional metrics for each event, covering critical path, wait states, (total) blame and critical blame. This enriched trace can be visualized using Vampir for further manual inspection and analysis. Vampir features three options to visualize the included metrics[1]: The *counter data timeline* plots the value of any metric for a specific event stream over time. The *performance radar* allows the user to inspect a metric's values over time for all event streams simultaneously using color coding. The third option is the *performance overlay* mode where a color-coded version of any metric is displayed transparently over the original activity timeline.

In addition to counters added to the trace by CASITA, Vampir provides the ability to create user-defined *customized metrics*. In their most simple application, they can be used to create a base-ten logarithmic version of each initial counter. This is helpful because the value range of aggregated waiting time or blame per activity can be very large, spanning multiple orders of magnitude. Creating a logarithmic derived metric allows to visualize the whole value range more easily using color coding. However, re-writing the original trace to include the raw analysis results has several limitations. The trace size grows significantly as for every event of any analyzed programming model, one or multiple performance counters must be additionally stored. For large-scale traces, simply storing the trace a second time on disk can therefor require an unpleasant amount of the tool's runtime. Section 5 therefore addresses strategies for reducing this overhead in future work.

3.4   Use Cases

In the following, the applicability of CASITA's critical-blame analysis capabilities and its integration into the Score-P/Vampir measurement and visualization workflow is demonstrated using two sample programs. As examples, we use a multi-paradigm implementation of the Jacobi method as well as a hybrid MPI+OpenMP version of LULESH.

3.4.1   Jacobi Method

The chosen implementation of the Jacobi method exploits MPI, OpenMP and CUDA [2]. At benchmark start, one can specify the ratio of work that will be offloaded to the CUDA device. The remainder is computed using OpenMP on the available host CPU threads. Figure 4 shows a scenario where the workload is almost equally balanced between host and CUDA device. Neither the CPU nor the GPU is dominating the critical path. However, the performance radar on the bottom of Fig. 4 shows that the *OpenMP threads 1:1* and *1:0* are blamed for the waiting time they are causing in other OpenMP threads of their team, which reveals a load-imbalance between the OpenMP threads. The hot spot analysis clearly identifies an OpenMP for loop on line 177 as runtime dominating. However, optimizing this OpenMP region would only result in additional waiting time on the host (for the

---

[1] The terms *metric* and *counter* are used interchangeably by Vampir.

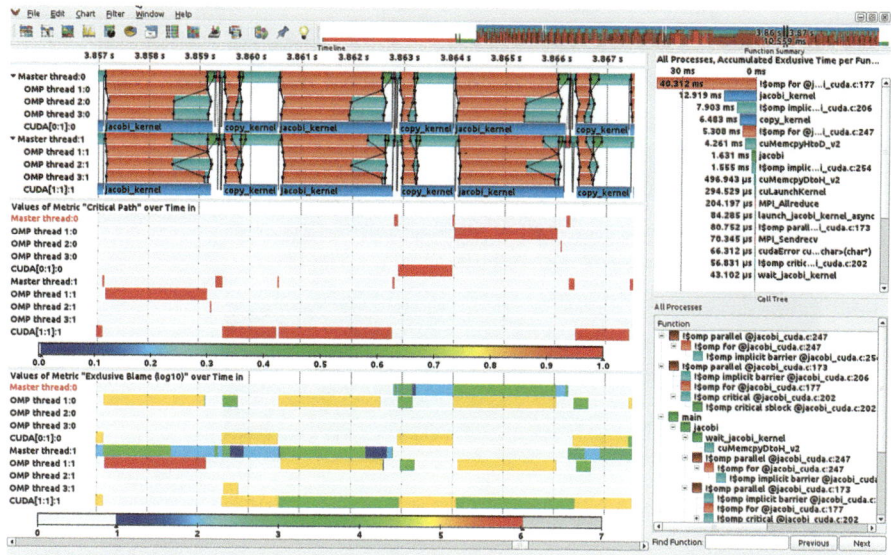

Fig. 4 Jacobi Method: Vampir screenshot for a scenario where the workload is almost equally balanced between CPU and CUDA device. The *Master Timeline* (*top display*) distinguishes activities according to the instrumentation. The *Function Summary* (*top right display*) shows the results of the hot-spot analysis (exclusive runtime) for each activity type. The *Performance Radar* (*middle display*) exposes the critical path (*red*), which changes between all available execution streams. The *bottom Performance Radar* uses color coding to illustrate the blame distribution

GPU). The critical-blame analysis identifies both CUDA kernels (*jacobi_kernel* and *copy_kernel*) as most viable optimization candidates. These are the correct optimization targets to reduce the total program runtime, which can be seen in the figure for the shorter running *copy_kernel* that causes the CPU to be unused for a period of time (white areas in the *Master Timeline* on the top). Note that the critical-blame metric used in the optimization rating is calculated for the full trace, whereas zooming in Vampir enables the user to analyze arbitrary time ranges.

3.4.2 LULESH

The second use case is based on the popular LULESH benchmark [11], a proxy code representing a real-world three-dimensional Lagrangian hydrodynamics simulation software. The application implements MPI and OpenMP and features the option to induce artificial load imbalances between MPI ranks. Figure 5 visualizes an OTF2 trace with additional counter data generated by CASITA. It shows two representative iterations of the code that has been executed with eight MPI processes each spawning two OpenMP threads. The hot-spot analysis points towards the most runtime consuming function *omp@lulesh.cc:810* (topmost entry, right display).

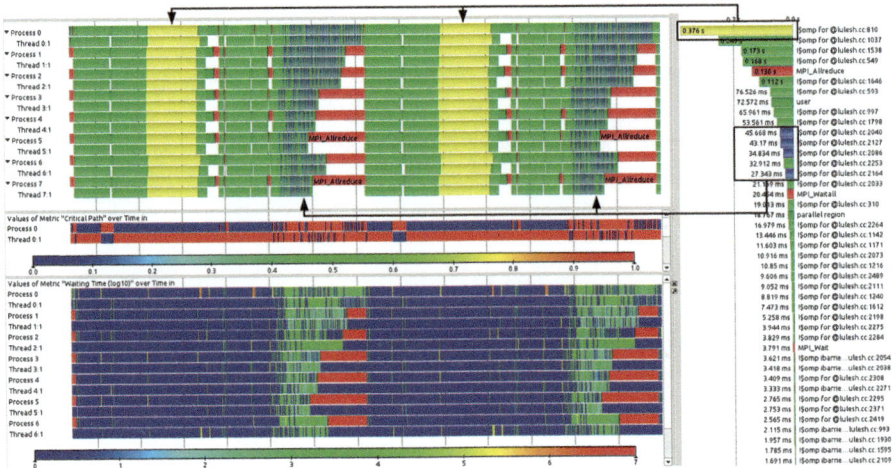

Fig. 5 LULESH: Two representative iterations visualized with Vampir. The additional counter data generated by CASITA enable a detailed visualization of hybrid load-imbalance problems. Typical hot-spot analysis is based on the exclusive runtime (*right display*) and therefore highlights the longest running activity (*yellow*). However, its execution is very balanced (*top display*). In contrast, other OpenMP loops (*blue*) induce load-imbalances and therewith waiting time between processes and threads (*bottom display*). At the `MPI_Barrier`, all processes must wait on *Process 0* which is detected as the critical path (*center display*)

However, critical-blame analysis reveals four other OpenMP loops (blue) as more important optimization candidates. Although *omp@lulesh.cc:810* has a higher total exclusive runtime, it is balanced across all threads and processes. The OpenMP loops marked blue in the figure induce much more waiting time (bottom display) for which they are attributed higher blame values. Hence, CASITA is able to accurately identify wait states and their root causes for hybrid MPI/OpenMP applications. The global load-imbalance in the execution results in all processes being stalled waiting on *Process 0* in the MPI_Allreduce (red) at the end of each iteration. As a result, the critical path is solely on Process 0 and *OpenMP Thread 0:1* spawned by this process (see center timeline display).

4 Related Work

The critical path is an important property of parallel programs. It has been widely discussed in literature, especially for message-passing programming paradigms. Schulz [19] describes a method to extract the critical path from MPI applications as a graph. Boehme et al. [3] present a scalable approach to compute performance metrics based on the critical path of an MPI trace. They implemented it as part of the commonly-known Scalasca toolset [10]. For example, their *critical-path imbalance indicator* determines the load-imbalance in an activity that executes across multiple processes with respect to the time it is contributing to the program's critical path. Similarly to CASITA, Scalasca is based on the Score-P measurement infrastructure [15], using both CUBE profiles as well as OTF2 trace files as input formats.

Additionally, Scalasca can perform a root-cause analysis on traces of MPI and OpenMP programs and to generate call-path profiles. As shown by Boehme et al. in [4], it is able to identify direct waiting time, occurring as a result of synchronization or communication inefficiencies along with indirect wait states. The latter arise from a propagation of direct inefficiencies through the program execution. Subsequently to the wait-state analysis, the program is replayed backwards, beginning from the last `MPI_Finalize` call, to locate the critical path of the MPI execution. This approach scales very well since the analysis runs as many processes as the analyzed program and the original communication pattern is re-enacted. This parallel reverse replay has been adopted by CASITA for its MPI critical-path analysis stage, too. The analysis report generated by Scalasca can be investigated with CUBE as well as other third-party tools such as ParaProf [18]. With respect to result visualization, Scalasca is not able to present generated information along a timeline and across multiple processes and threads.

HPCToolkit [1] consists of a set of tools for various aspects of performance data collection, analysis and visualization. It uses blame shifting [22] to attribute a penalty cost to the root cause or possible suspects of a wait state in a parallel application. This is possible for both MPI and OpenMP codes but can also be used to investigate contention problems in locking-based models [13]. HPCToolkit can apply its blame-shifting approach to CUDA programs by attributing blame to CUDA kernels on the device that force the host execution to wait on their completion or vice versa [5]. However, it can not analyze waiting time or distribute blame occurring as a result of intra-device dependencies. Unlike Score-P, HPCToolkit collects data using sampling instead of instrumentation. For every sampling point, the active call path along with further performance metrics is captured and combined in a calling-context tree. Idle regions such as wait states are detected by identifying blocking routines by name. HPCToolkit is able to blame certain activities for generating waiting time, however, it does not provide insight on functions that are critical for the global program runtime. Additionally, it can not apply wait-state analysis to codes that use a combination of all three levels of parallelism (CUDA+OpenMP+MPI) and for OpenMP analysis, an altered version of the GNU OpenMP runtime is necessary.

A reasonable visualization of program traces facilitates the detection of a performance problem which has been shown in [12]. Nevertheless, it is hard to identify potential optimization candidates that determine the global runtime by investigating process time lines and applying hot spot analysis, only. This work adds the possibility to highlight the critical path and blame in the Vampir time line view, which guides the application developer to optimization spots that are most relevant for global program runtime reduction.

5   Conclusion and Future Work

This work presented our integration of CASITA's hierarchical critical-blame analysis for multiple programming models into the Score-P performance analysis environment. During measurement, the user is only required to set an additional environment variable flag to enable dependency tracking. Afterwards, Score-P's OTF2 trace output can be used directly as input to CASITA's scalable analysis. The generated optimization guidance profile aids the user in quickly identifying valuable optimization targets based on the critical-blame performance metric. Additionally, CASITA can create an enriched OTF2 trace including various performance-relevant metrics such as the critical path, waiting time and distributed blame. Modifications to input and output traces are purely based on the OTF2 standard and are thus non-intrusive to other tools in the Score-P workflow. Analysis results in the re-written trace can be manually investigated using the trace browser Vampir.

There are several issues and aspects that will be addressed in the future development of CASITA. Although many traces could be successfully analysed we are aware that in a few cases the analysis still fails or does not provide correct results. Therefore, the program stability will be addressed by implementing automated testing. To improve the user experience it has to be integrated into a stable Score-P release and the resulting software package. Second, we are investigating options to integrate CASITA's analysis results more tightly into Vampir. The primary challenge is the costly trace re-write, in terms of required time and disk size. Extending the OTF2 library to enable adding counter data streams in an existing OTF2 trace will address both issues. A closer integration into Vampir would also enable new visualization approaches. For example a new general profile display could be used to present the optimization guidance report.

References

1. Adhianto, L., Banerjee, S., Fagan, M., Krentel, M., Marin, G., Mellor-Crummey, J., Tallent, N.R.: HPCTOOLKIT: tools for performance analysis of optimized parallel programs. Concurr. Comput.: Pract. Exp. **22**(6), 685–701 (2010). doi:10.1002/cpe.1553
2. Adinetz, A., et al.: NVIDIA application lab at JSC. In: inSiDE, vol. 11 (2013)
3. Bohme, D., Wolf, F., De Supinski, B., Schulz, M., Geimer, M.: Scalable critical-path based performance analysis. In: 2012 IEEE 26th International Parallel Distributed Processing Symposium (IPDPS), Shanghai, pp. 1330–1340 (2012). doi:10.1109/IPDPS.2012.120
4. Bohme, D., Wolf, F., Geimer, M.: Characterizing load and communication imbalance in large-scale parallel applications. In: 2012 IEEE 26th International Parallel and Distributed Processing Symposium Workshops PhD Forum (IPDPSW), Shanghai, pp. 2538–2541 (2012)
5. Chabbi, M., Murthy, K., Fagan, M., Mellor-Crummey, J.: HPCToolkit: a tool for performance analysis on heterogeneous supercomputers. In: GTC 2013, San Jose (2013)
6. Dietrich, R., Schmitt, F., Grund, A., Schmidl, D.: Performance measurement for the OpenMP 4.0 offloading model. In: Lopes, L. (ed.) Euro-Par 2014: Parallel Processing Workshops, Porto, vols. 8805, 8806 (2014)

7. Dietrich, R., Schmitt, F., Grund, A., Stolle, J.: Critical-blame analysis for OpenMP 4.0 offloading on Intel Xeon Phi. In: First Workshop on Software Engineering for Parallel Systems (SEPS), Portland (2014)
8. Eichenberger, A., Mellor-Crummey, J., Schulz, M., Copty, N., Cownie, J., Dietrich, R., Liu, X., Loh, E., Lorenz, D.: OpenMP technical report 2 on the OMPT interface. Technical report (2014)
9. Eschweiler, D., Wagner, M., Geimer, M., Knüpfer, A., Nagel, W.E., Wolf, F.: Open trace format 2 – the next generation of scalable trace formats and support libraries. In: Proceedings of the International Conference on Parallel Computing (ParCo), Ghent, 30 Aug–2 Sept 2011. Advances in Parallel Computing, vol. 22, pp. 481–490. IOS Press (2012)
10. Geimer, M., Wolf, F., Wylie, B.J.N., Erika Abraham, D.B., Mohr, B.: The Scalasca performance toolset architecture. Concurr. Comput.: Pract. Exp. **22**(6), 702–719 (2010)
11. Karlin, I., Keasler, J., Neely, R.: Lulesh 2.0 updates and changes. Technical report LLNL-TR-641973 (2013)
12. Knüpfer, A., Brunst, H., Doleschal, J., Jurenz, M., Lieber, M., Mickler, H., Müller, M.S., Nagel, W.E.: The Vampir performance analysis tool-set. In: Resch, M., Keller, R., Himmler, V., Krammer, B., Schulz, A. (eds.) Tools for High Performance Computing, Proceedings of the 2nd International Workshop on Parallel Tools for High Performance Computing, Stuttgart. Springer, Berlin/Heidelberg (2008)
13. Liu, X., Mellor-Crummey, J., Fagan, M.: A new approach for performance analysis of OpenMP programs. In: Proceedings of the 27th International ACM Conference on International Conference on Supercomputing, Eugene, pp. 69–80. ACM (2013)
14. Message Passing Interface Forum: MPI: a message-passing interface standard, version 2.2 (2009)
15. Mey, D., Biersdorf, S., Bischof, C., Diethelm, K., Eschweiler, D., Gerndt, M., Knüpfer, A., Lorenz, D., Malony, A., Nagel, W.E., Oleynik, Y., Rössel, C., Saviankou, P., Schmidl, D., Shende, S., Wagner, M., Wesarg, B., Wolf, F.: Score-P: a unified performance measurement system for petascale applications. In: Bischof, C., Hegering, H.G., Nagel, W.E., Wittum, G. (eds.) Competence in High Performance Computing 2010, pp. 85–97. Springer, Berlin/Heidelberg (2012)
16. Mohr, B., Malony, A.D., Shende, S., Wolf, F.: Design and prototype of a performance tool interface for OpenMP. J. Supercomput. **23**(1), 105–128 (2002). doi:10.1023/A:1015741304337
17. NVIDIA: CUDA Toolkit Documentation – CUPTI. http://docs.nvidia.com/cuda/cupti/index.html (2013)
18. Performance Research Lab: ParaProf user's manual. University of Oregon. http://www.cs.uoregon.edu/research/tau/docs/paraprof/ (2010)
19. Schulz, M.: Extracting critical path graphs from MPI applications. In: IEEE International Cluster Computing, 2005, Boston, pp. 1–10 (2005). doi:10.1109/CLUSTR.2005.347035
20. Schmitt, F., Dietrich, R., Juckeland, G.: Scalable critical path analysis for hybrid MPI-CUDA applications. In: 2014 IEEE 28th International Parallel and Distributed Processing Symposium AsHES Workshop (IPDPSW), Phoenix (2014)
21. Schmitt, F., Stolle, J., Dietrich, R.: CASITA: a tool for identifying critical optimization targets in distributed heterogeneous applications. In: 2014 44th International Conference on Parallel Processing Workshops (ICPPW) (2014)
22. Tallent, N., Adhianto, L., Mellor-Crummey, J.: Scalable identification of load imbalance in parallel executions using call path profiles. In: 2010 International Conference for High Performance Computing, Networking, Storage and Analysis (SC), New Orleans, pp. 1–11 (2010). doi:10.1109/SC.2010.47
23. TOP 500 supercomputer Sites. http://top500.org/lists/2014/06/ (2014). 43rd List
24. Yang, C.Q., Miller, B.: Critical path analysis for the execution of parallel and distributed programs. In: 8th International Conference on Distributed Computing Systems, 1988, San Jose, pp. 366–373 (1988). doi:10.1109/DCS.1988.12538

Studying Performance Changes with Tracking Analysis

Germán Llort, Harald Servat, Juan Gonzalez, Judit Gimenez,
and Jesús Labarta

Abstract Scientific applications can have so many parameters, possible usage scenarios and target architectures, that a single experiment is often not enough for an effective analysis that gets sound understanding of their performance behavior. Different software and hardware settings may have a strong impact on the results, but trying and measuring in detail even just a few possible combinations to decide which configuration is better, rapidly floods the user with excessive amounts of information to compare.

In this chapter we introduce a novel methodology for performance analysis based on object tracking techniques. The most compute-intensive parts of the program are automatically identified via cluster analysis, and then we track the evolution of these regions across different experiments to see how the behavior of the program changes with respect to the varying settings and over time. This methodology addresses an important problem in HPC performance analysis, where the volume of data that can be collected expands rapidly in a potentially high dimensional space of performance metrics, and we are able to manage this complexity and identify coarse properties that change when parameters are varied to target tuning and more detailed performance studies.

1 Background and Motivation

The execution of a scientific code is dependent on a variety of parameters that may have a strong impact on its performance. Some examples include the size of the input problem, the number of processes running in parallel, the physical mapping and sharing of the resources, the parallel programming model used, and many other settings. Anticipating the impact of different configurations on the achieved performance, work balancing or memory usage of the program is far from trivial and not seldom leads to discover unexpected issues.

G. Llort (✉) • H. Servat • J. Gonzalez • J. Gimenez • J. Labarta
Barcelona Supercomputing Center – Polytechnic University of Catalonia – BarcelonaTech, Jordi
Girona 31, 08034, Barcelona, Spain
e-mail: gllort@bsc.es; harald@bsc.es; jgonzale@bsc.es; judit@bsc.es; jesus@bsc.es

© Springer International Publishing Switzerland 2015
C. Niethammer et al. (eds.), *Tools for High Performance Computing 2014*,
DOI 10.1007/978-3-319-16012-2_9

Analyzing these effects is important not only to get better understanding of the program behavior, but also to foresee improving or degrading trends in the different parts of the code, identify the main limiting factors, and in the end, to help the users making the right decisions to tune the application to achieve the most performace outcome. To this end, it is necessary to have tools to easily compare different experiments and correlate observations between them.

In order to deal with the difficulties inherent to running, measuring and comparing multiple experiments, we have designed a tool to conduct very diverse parametric and evolutionary studies, enabling to correlate performance information either from multiple runs with different configurations, or different time intervals within the same experiment. Our approach focuses on the computational behavior of the most relevant code regions and shows their evolution with respect to several performance metrics to explain which factors lead the different parts of the code to improve or degrade. In this context, object tracking techniques become a natural and intuitive way to detect the performance changes sustained by each part of the code automatically, and represent the information in a clear and visual manner.

While previous approaches for comparing experiments or phases [12, 25, 26] have been proposed, our work goes one step further and presents a novel technique that does not rely on preselected metrics and profile data for static code phases, such as routines, loops or user-defined sections. One problem of summarizing the data at these levels is that one same section of code can exhibit behavior variations, thus making averages will hide divergent performance trends. Our position is that it is necessary not to consider averages, but every independent instance to detect fine-grain structure and capture multi-modal variability.

2   Object Tracking for Performance Analysis

Tracking techniques have been traditionally used to follow moving objects in an image or video sequence. Practical examples include augmented reality, medical imaging, surveillance or traffic control. A first step to these problems is to delimit the objects of interest within the scene depicted in the image. Therefore, object recognition algorithms (e.g. image segmentation and edge detection) will look for appearance characteristics and distinguishing features (e.g. color, direction or shape) that identify them. Then, consecutive frames in the sequence are compared to find correspondences between the objects and their displacements.

Analogously, we represent different executions as images, each one picturing the program behavior for a given configuration, and arrange them as a sequence of images that expresses the evolution of the application behavior across experiments. Code regions are drawn in the images as independent trackable objects, in a space whose dimensions are not the actual physical dimensions of height, length and breadth, but performance metrics that describe how these regions behave. Movements in the performance space across the images highlight changes in the application behavior, that can be modeled into metrics to evaluate the performance trends of the different regions of code.

This approach is useful to discover valuable performance insights about the application response to different configurations, enabling the analyst to draw quick conclusions on the key factors limiting performance, direct the optimization effort and easily determine the best setup to maximize a certain performance requirement. Throughout this chapter, we will be showing how this method applies to very diverse cases of analysis to get better understanding of the impact of different architectures, input problems, workloads, memory and resource sharing schemes, and levels of scalability on several parallel programs.

2.1 Application Structure Characterization

Analysis tools usually display performance data to the user in the form of profiles at the level of syntactic program structures (i.e. subroutines, loops, or user-defined sections). This has the advantage of providing a very natural and understandable representation, but also carries some drawbacks along. Prior knowledge of the application may be required to determine which functions are relevant, so as to skip too fine-grain routines that would perturb the execution due to the instrumentation overhead. When no automatic interposition mechanisms are available [4], access to the sources and manual modifications are needed to inject measurement probes in these points of interest. Moreover, considering a whole routine as a single unit of behavior can be deceitful, because different invocations may behave differently, depending on the parameters and conditional phases leading to distinct code flows with divergent performance. In these cases, a global average may convey the wrong idea of a reasonable overall behavior, while specific sub-phases may be reporting low performance and their optimization could lead to significant improvements, as proven in [23, 24].

A different granularity to characterize the application performance is the computing regions (i.e. CPU bursts). These are defined as the sequential computations between calls to the MPI or OpenMP runtime. Delimiting these regions only requires library interposition to instrument the parallel programming API, thus there is no need for user intervention nor access to the sources. Each CPU burst is described by its duration, call stack references that point to the corresponding source code, and a vector of hardware counters metrics describing how it performed. Considering every CPU burst rather than simple averages, we can detect variabilities across processes and time, exposing a fine-level characterization of every code region and the nature of their inefficiencies.

This approach is less attached to the structure of the source code, but focuses on the performance properties of the actual computations. In [9], the authors prove that this granularity is useful for the analysis of parallel programs, as it reflects an intermediate point of view between very low level characterizations (i.e. basic blocks or instruction-level simulators) and higher abstractions (i.e. functions, loops or user-defined sections). Regardless of our implementation, which selects CPU bursts as the target granularity, the technique presented would as well be applicable using other abstractions.

2.2 Generation of Tracking Images

In computer vision, one or more particular objects (e.g. humans, cells or cars) are first identified within a frame (a single picture in a video or series of images) and then tracked as they move through a sequence of frames. Likewise, we are going to identify the computing regions of interest and keep track on how their performance evolves along multiple experiments. To this end, we first need to represent the performance measurements observed in each experiment graphically, or in other words, to capture our sequence of frames. This process consists in selecting any pair of metrics to draw a two-dimensional space where we express the behavior of every individual CPU burst with a point in the plane. Typically, we select *Instructions per Cycle (IPC)* and *Instructions Completed*, which are useful to bring insight into the overall performance: trends in *Instructions Completed* indicate regions with different workloads, while *IPC* measures how fast the work is done. Anyhow, this process can be applied to any arbitrary combination of metrics that may be used to describe the CPU bursts (e.g. cache misses, floating-point operations or power consumption) to support even more precise multi-dimensional characterizations of the data.

With the images generated, the next step is to identify the objects of interest within them. Due to the highly iterative nature of HPC applications, many computations will be very alike in terms of the performance they achieve. In the image, this translates as clouds of points that are close in the space, which can be grouped into a single entity according to their similitude. Therefore, we apply density-based cluster analysis [9, 11] in order to group similar CPU bursts with respect to the metrics selected.

The result of this process is a scatter-plot representation of the performance space, where the axes correspond to the metrics used to cluster the data, and all CPU bursts that are similar with respect to these metrics get grouped into the same object. Clusters are then intrinsically connected to the source code regions of their belonging CPU bursts, and both terms will be indistinctly used for clarity, but this connection is not necessarily unambiguous: a single region presenting bimodal behavior will result in two distinct clusters, while two different regions with similar behavior will conform the same cluster. So in essence, what each cluster represents is a behavioral trend, independently of the code region that exhibits it.

One question that may arise about the benefits of using these performance images is to what extent they are better than just a straightforward profile. To dispel the doubt, we have selected as example the BT-MZ benchmark [18], a solver for block tri-diagonal systems that performs computations of uneven size. Table 1 shows the average IPC, total instructions and L1 misses scored by three of the main functions,

Table 1 User functions
profile for BT-MZ

	IPC	Instructions	L1 miss
x_solve	2.16	43.04 M	295.92 K
y_solve	2.16	43.83 M	323.07 K
z_solve	2.17	46.22 M	55.63 K

Fig. 1 Clusters for 3 main
functions of BT-MZ (Class B,
4 tasks)

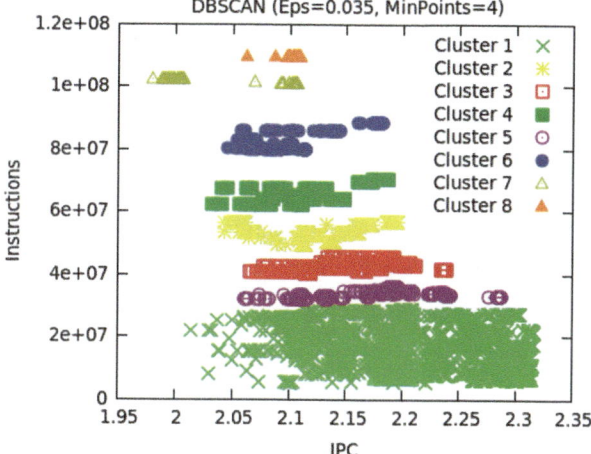

measurements obtained by instrumenting the routines at their start and end points.
From these numbers, we can easily infer that all three routines present a similar
computational behavior, with the same amount of work (Instructions) executed at
the same speed (IPC), yet they show different memory efficiency with lower L1
cache misses in the Z-direction, certainly due to the data access pattern. One could
expect this result, as these functions perform the same kind of computation over
different axes.

Figure 1 shows the performance image generated for these functions, with each
point in the plot being a single instance of invocation, and grouped in clusters with
respect to the IPC achieved and the number of instructions executed. A function-
agnostic view of the data brings new insights about the application structure:
all three functions show eight different computational behaviors with increasing
amounts of work and decreasing speed. Computations with high amount of work but
low performance are interesting to study, as well as those with the same amount of
work at different speeds, or vice-versa, as these indicate potential load-imbalances.
All eight behaviors are exhibited by all three functions, which still conveys the idea
that these functions are similar, but exposes their inner variability as they behave
more or less optimal depending on the size of the workload.

Table 2 shows the same statistics for the clusters, and now you can easily see
a large dynamic range in the metrics. Most significantly, a standard deviation of
30 M of instructions reveals a large work imbalance between all clusters, which was
masked in a traditional function-based profile. Column *% Time* shows the fraction
of the total execution time that these computational behaviors cover, and it is clear

Table 2 Clusters profile for BT-MZ

	IPC	Instructions	L1 miss	% time
Region 1	2.21	19.15 M	56.45 K	36.95 %
Region 2	2.13	53.46 M	266.33 K	12.28 %
Region 3	2.16	42.36 M	194.32 K	12.08 %
Region 4	2.12	65.79 M	363.01 K	11.43 %
Region 5	2.18	33.87 M	133.19 K	11.42 %
Region 6	2.11	83.27 M	494.41 K	9.68 %
Region 7	2.05	101.61 M	949.55 K	4.01 %
Region 8	2.10	109.46 M	115.66 K	2.13 %

that their weight is not negligible, and thus the importance of being aware of these variabilities. This example highlights the importance of focusing on the dynamic behavior of the regions rather than static code structures to guarantee that we detect performance variabilities and direct the analysis towards the zones of real interest.

2.3 Tracking Difficulties

The main difficulty in the use of tracking techniques arises due to abrupt object motions and noise in the images. When applied to performance analysis the problem is the same. Even though one would normally expect the application performance not to radically change all of a sudden, performance variations may result in large changes of behavior, preventing us from borrowing any assumption about the clusters' position, direction or shape in the performance space.

The clustering process of a frame assigns numbers and colors to every cluster identified. Since this is an independent, non-supervised process, the clustering of a second, different frame does not necessarily have to result in the same number of objects, assign the same identifiers, or exist a direct correspondence between their numberings. Figure 2 shows the structure of the 12 most time-consuming regions of WRF [29] ran with 128 processes. Clusters are formed according to similarities in the achieved performance (X-axis) and number of instructions (Y-axis). Those that stretch vertically (e.g. Region 2) denote instructions imbalance, while those that stretch horizontally (e.g. 7 and 11) reflect IPC variations. Figure 2b shows the structure of WRF doubling the number of cores. The number of instructions executed per core has reduced in inverse proportion, and so all clusters have moved downwards the Y-axis. Intuitively, we can see that cluster 2 (yellow) turned into 3 (red). And a few clusters have slightly improved their performance (e.g. 4 and 6 moved right with higher IPC), while cluster 11 significantly degraded. But some changes are far from evident: zooming into the boxed areas, you can see a fourth cluster appearing. Is that the left-most cluster in the 128-task case redistributed into the two small ones on the left of the 256-task case? Or these two come from split parts of the two left-most clusters?

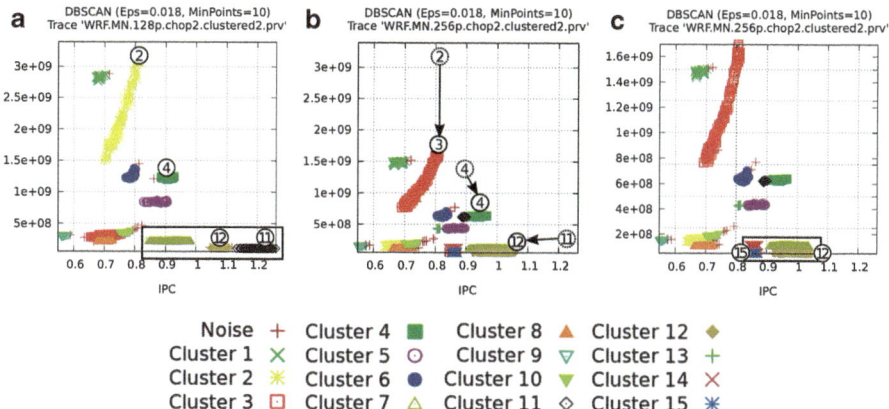

Fig. 2 Structure of WRF computing bursts. (**a**) 128 tasks. (**b**) 256 tasks. (**c**) 256 tasks normalized

With changing scenarios that may affect the application performance, clusters can not only move long distances or change their shape between frames, they can also vary in density, split, or merge together. And if the configurations that differentiate the experiments vary significantly, the frames to compare can be remarkably different, which makes even more difficult to detect the interesting regions and see how they change from one frame to the next. Although in some cases it would be possible to determine who-is-who by visual inspection, this is not obvious in the general case, and so the benefits of an automated mechanism able to detect abrupt changes amongst many clusters become palpable.

The first difficulty in determining which objects within a frame correspond to the ones in the next lies on the fact that the respective scales may be different, so they can not be compared directly. For example in a strong-scaling case, when the number of cores increases, the number of instructions executed per core will decrease in proportion. A step prior to track the evolution of the objects consists in normalizing the performance scales so that they are comparable. Such metrics that are correlated with the number of processes of the application (e.g. Instructions) are weighted by the number of cores, while the scale for the rest (e.g. IPC) is adjusted to the minimum and maximum values seen along all experiments. Figure 2c shows the 256-tasks case with the performance scales normalized. The relative distances compared to the base 128-tasks case are kept almost constant, and the experiments can now be easily compared.

In the next section we present a tracking algorithm that performs an automatic correlation of equivalent code regions that are subject to performance variations along multiple experiments. To this end, we extrapolate the concept of recognizing moving objects in a sequence of images to the displacement of clusters within the metrics space across experiments. Clustering the application performance can be seen as identifying the objects of interest (regions of code with a certain behavior) in a single frame. Subsequent clusterings result in a sequence of images that can be compared to see how these objects move, shape-shift, merge or split in the performance space, reflecting changes in the application behavior. Tracking their evolution across experiments enables us to study the performance characteristics of the different code regions, and to understand how the different configurations get to influence their behavior.

2.4 Implementation Details

The current implementation uses the Extrae tracing toolkit [3] to automatically instrument MPI and/or OpenMP codes through library preloading techniques. For each entry and exit point of the parallel runtime, the tool writes a per-thread timestamped event trace, and collects hardware counters data through PAPI [1], and source code references by using libunwind [28] to walk the call stack and GNU binutils [8] to fetch human-readable debugging information from the binary. In our experiments, the size of the traces generated ranged from tens of MB to tens of GB.

The clustering tool extracts the CPU bursts data comprised in the trace and runs a basic DBSCAN algorithm to identify the main computing trends. In this process, bursts with very short duration are considered negligible and discarded, so as to avoid the high cost of processing many small points. In [9], the authors prove that one can discard up to 80 % of the data, while preserving the 99 % of the computation representativity. This clustering tool can process up to 100 K points under 1–2 min.

As reported in the literature, tracing tools already scale to hundreds of thousands of cores [7], and parallel density-based algorithms are able to manage millions of points [21]. Once the data has been reduced to representative clusters in the performance image, the tracking algorithm presented next works with a very reduced number of objects, enabling low response times from few seconds to few minutes, and so the technique presented, relying on large-scale tracing and clustering tools, is perfectly applicable with large volumes of data and totally scalable.

3   The Tracking Algorithm

The objective of this algorithm is to automatically correlate equivalent computational components that are subject to performance variations, tracking how they move along a sequence of images that represent the application's performance behavior. Let A and B be two images, as depicted in Fig. 3, where n and m objects are respectively detected, say $A = \{A_1, A_2, \ldots, A_n\}$ and $B = \{B_1, B_2, \ldots, B_m\}$. The objective is to find the maximum number of relations k, so that exists a k-partition $P = \{P_1, \ldots, P_k\}$ of A, and a k-partition $Q = \{Q_1, \ldots, Q_k\}$ of B, that fulfill the condition:

$$\forall i : 1 \leq i \leq k : P_i \equiv Q_i$$

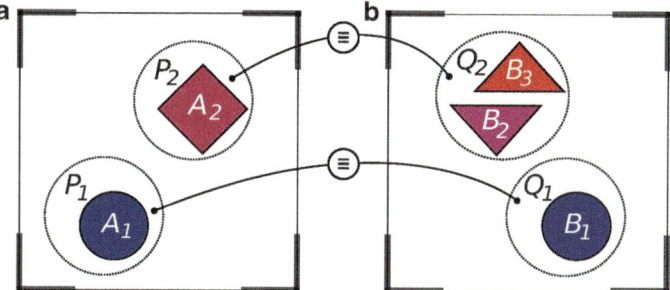

Fig. 3 Tracking scheme

Where the optimal k is bounded above by the image with the fewer number of objects detected, i.e. $min(n, m)$, and the equivalence relation $P_i \equiv Q_i$ is the assumption that objects in partition P_i correspond to those in partition Q_i.

In order to determine whether two clusters are equivalent, there are three principal properties of the computations that can be considered: the position in the performance image, the position in the source code, and the position in the execution trace. Based on these characteristics, we define five complementary heuristics to evaluate the clusters equivalences that are detailed in the next section.

3.1   The Tracking Heuristics

Recalling the difficulties to apply tracking on performance data that we previously discussed in Sect. 2.3, deciding whether two clusters from different experiments represent the same computational behavior requires to consider several characteristics of the computations. In our implementation, each characteristic is evaluated with a different heuristic. Applying just a single heuristic is generally not enough, because as we will discuss throughout this section, most of the characteristics inspected are to some extent ambiguous and do not allow to perfectly differentiate between the objects. Moreover, not all the information required to apply all the heuristics is always present (that depends on the system and the amount of information collected during the instrumentation phase). Therefore, we employ multiple heuristics and combine their results to decide the equivalences between all objects. Each heuristic focuses on a particular characteristic of the computations:

- *Distance of the movement.* Clusters can move in any direction of the space as a consequence of performance variations, but in the general case, these will manifest as smooth, directed transitions rather than swift leaps. For example, if we keep increasing the size of the workload, we can expect the total number of instructions executed in all computations to increase as well, and make certain assumptions on the directions of the movements.

- *SPMDiness.* In SPMD applications, all processes must be executing the same code phase simultaneously. If two different clusters happen at the same time, since the application is SPMD they can not refer to different code phases, and so they must be the same code phase that is presenting multi-modal behavior.
- *Call stack references.* Call stack information links every CPU burst in a cluster to the function, file and line in the source code where it is executed. Different clusters can not be the equivalent if the computations that form them do not share any call stack reference to the same point in the source code.
- *Clusters density.* If there are performance variabilities that make the clusters split, there must be a combination of the split clusters so that the sum the computations that form each cluster equals the total number of computations that form the equivalent unsplit cluster in the previous experiment.
- *Chronological sequence.* Two experiments running the same program will show the same time-ordered sequence of computations, so those that appear in the same order of occurrence must be equivalent.

The following Sects. 3.1.1–3.1.5 describe each of the heuristics in more detail. Then in Sect. 3.2 we explain how the information provided by the different heuristics is combined to maximize the number of objects successfully tracked.

3.1.1 Distance of the Movement

This heuristic takes a pair of images and performs a cross-classification of every computing burst from the first into the latter, and vice versa. The classification is based on a nearest-neighbor criteria, so that all points will get classified to the nearest counterpart cluster. This can be seen as projecting each object from one image to the next, and see which object in the second image is closer.

The idea that lies behind supports on the fact that the behavior of a parallel application will not radically change along images, and so the objects displacements will generally be short. This assumes a certain ordering in the pairs of images that are compared, as the more different they are, the more difficult becomes to find correspondences. However, for the majority of analyses an implicit order emerges. Consider again the previous example where we doubled from 128 to 256 the number of cores in WRF (see Fig. 2a, c). The general structure for both experiments hardly differs, with very slight movements.

There are situations where a cluster may split into two or more. For example, when new zones of imbalance appear and separate one region into several distinct performance behaviors. This case can be seen in Fig. 4, where region A_4 shifts to two behaviors, namely B_4 and B_{11}. Also, there are cases where clusters can move a long way in the space, which is the case of regions 11 and 12 in Fig. 2a to regions 12 and 15 in Fig. 2c, respectively. In these situations, cross-classification based on distance is likely not to assign the points to the correct cluster (both get assigned to 12 because 15 is too far away, which illustrates a mapping error), but we can then use the next heuristics to discern whether those regions are the same or not.

	B_1	B_2	B_3	B_4	B_5	B_6	$\cdots$	B_{10}	B_{11}	B_{12}
A_1	100%	0	0	0	0	0		0	0	0
A_2	0	0	100%	0	0	0		0	0	0
A_3	0	99%	0	0	0	0		1%	0	0
A_4	0	0	0	34%	0	0		0	65%	0
A_5	0	0	0	0	100%	0		0	0	0
A_6	0	0	0	0	0	100%		0	0	0
$\vdots$										
A_{11}	0	0	0	0	0	0		0	0	100%
A_{12}	0	0	0	0	0	0		0	0	100%

Fig. 4 Cross-classification between WRF-128 and WRF-256

3.1.2 SPMDiness

This heuristic exploits the SPMD structure of the applications to match computing regions that happen simultaneously in different processes. Assuming this execution model, all processors are expected to be executing the same phase of code at a time. In this case, if multiple processes are executing different types of computations concurrently, they are likely to refer to the same code region, although there might be performance variations that make them shift apart (e.g. the application presents work imbalance).

Figure 5a shows a detailed view of the temporal sequence of clusters at the beginning of one iteration of WRF 128-tasks. All processes (Y-axis) execute the same computations over time (X-axis). The same pattern can be seen in Fig. 5b for the 256-tasks case, meaning that the code phases and the order in which they get executed are the same in both runs. However, in this case some processes are undergoing duration imbalances and execute longer computations, shown as stride lines with distinct colors. The new behavior is identified as a different cluster, but these are actually the same computing phases and can be linked together.

The application SPMDiness is evaluated with the technique presented in [10]. The algorithm takes as input the sequence of clusters for every task of the application, and performs a Multiple Sequence Alignment (MSA). Clusters from different tasks that fall into the same position of the globally aligned sequence are those that get executed simultaneously, and we use this information to mark them as equivalent. If the application follows a programming model that may result in different processes running different parts of the code at the same time (e.g. task-based parallelism), this heuristic alone may lead to inconclusive decisions.

3.1.3 Call Stack References

This heuristic prunes the search space by discarding matchings between regions that do not have call stack references in common. Call stack information points to

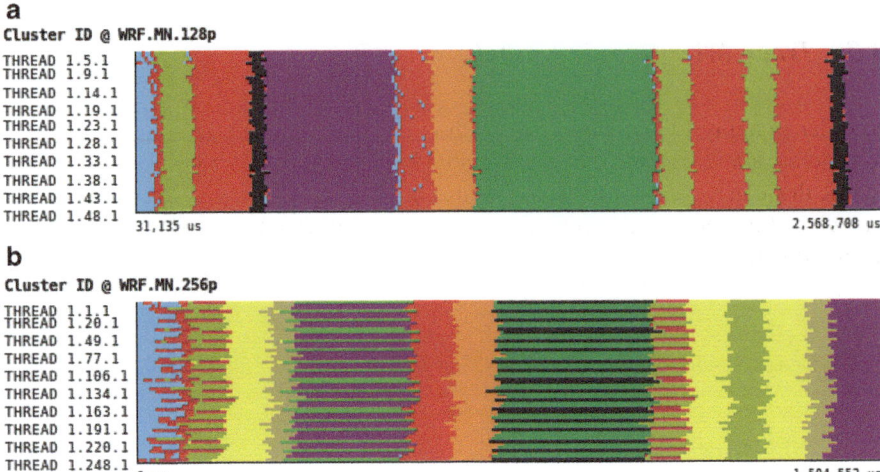

Fig. 5 Correlations from SPMDiness heuristic for WRF. (**a**) SPMD computations for WRF-128. (**b**) SPMD computations for WRF-256

Table 3 Correlations from call stack heuristic for WRF

128 tasks	Callstack references	256 tasks
Region 1	4939 (module_comm_dm.f90)	Region 1
Region 2	6474 (module_comm_dm.f90)	Region 3
		Region 5
Region 5		Region 13
Region 3	6060 (module_comm_dm.f90)	Region 2
Region 4	2472 (module_comm_dm.f90)	Region 4
		Region 11
Region 7	5734 (module_comm_dm.f90)	Region 7
Region 11	6275 (module_comm_dm.f90)	Region 12
Region 12		Region 15

the function, file and source code line where the CPU burst starts, linking them to specific points of code. If two clusters from two different frames do not share code references, they are certainly not equivalent.

Table 3 illustrates a subset of the relations that can be outlined between regions from their code references. The reason why some relations are ambiguous is because the clustering process groups computations based on their similarity with respect to the selected metrics to generate the performance images, so it is possible that different points of code behave the same and get grouped under the same cluster. Also, if a single code region presents multi-modal behaviors, it will appear as part of multiple clusters. This information alone is insufficient to discriminate more, but effectively reduces the combinatorial explosion.

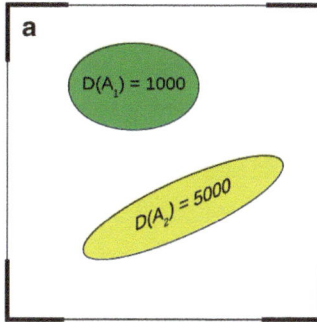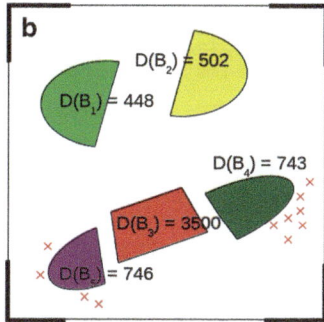

Fig. 6 Correlations from clusters density heuristic. The aggregate density of the split clusters on the *right* is lower or equal than the merged clusters on the *left*

3.1.4 Clusters Density

This heuristic is applicable when comparing experiments that have computed the same number of CPU bursts. In those cases, the aggregate of computations of all the clusters in each performance image will be the same. If the points distribution in the performance space does not change between experiments, the densities of the clusters will also be the same. When a cluster splits, two or more sub-clusters will have formed, and the sum of their densities will equal the density of the original super-cluster that contained them all, as illustrated in Fig. 6.

This problem can be formulated as a variant of the 0/1 knapsack problem [13]: given a cluster i in the experiment A with a certain density $D(A_i)$, find the combination of sub-clusters in the second experiment B that maximizes the sum of their densities $D(B_{sum}) = D(B_1) + D(B_2) + \ldots + D(B_N)$ so that the aggregate density $D(B_{sum})$ is lower or equal than the limit density $D(A_i)$.

3.1.5 Chronological Sequence

This heuristic assumes that the program execution order will not change between experiments, and so the sequence of computing bursts over time will preserve the same chronological order. If the execution flow of the program varies between experiments (e.g. the program is dependent on the input data set, and triggers different algorithms optimized for specific data sets), then this heuristic is not applicable. When the execution order is preserved, it is possible to determine equivalent code regions by looking into the position where the computations appear in the execution sequence and matching those in the same position.

The sequence alignment technique referred in [10] is applied now on two experiments, and we then compare the order of occurrence of the computations. For example, consider an experiment that executes a loop comprising four computing regions with different performance behavior, and so these get classified in four

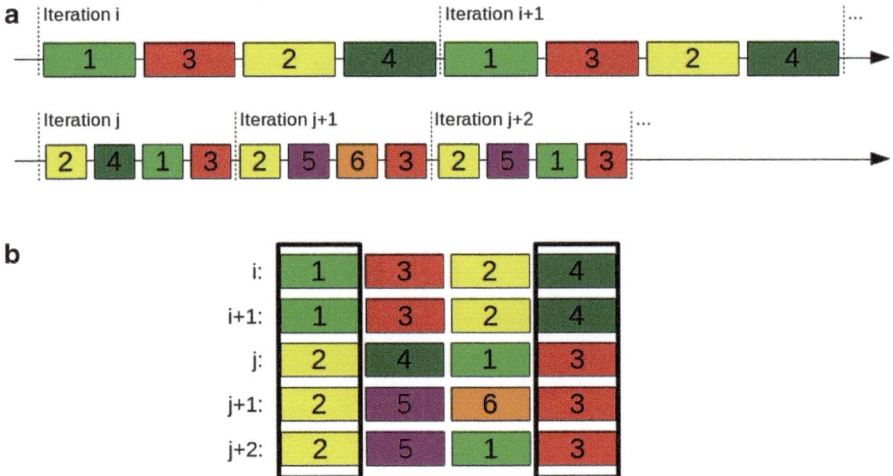

Fig. 7 Correlations from chronological sequence heuristic. (**a**) Sequence of computations in two different experiments. (**b**) Aligned subsequences between selected pivots, given that cluster 1 and 4 in the first sequence correspond to 2 and 3 in the second. Attending to their chronological order all clusters that fall the same column would be equivalent

different clusters. The top timeline in Fig. 7a depicts 2 iterations of this loop, with each computation colored according to the cluster to whom it belongs. A second experiment that uses more processes and a bigger problem size results in shorter computations and more iterations of the loop, as illustrated in the bottom timeline in Fig. 7a.

As we have discussed earlier in Sect. 2.3, the clustering process applied to different experiments can result in different clusters, hence having the same clusters colors or identifiers does not necessarily imply that they represent the same computing region, and so these sequences can not be compared directly. However, if we could guarantee some correspondences between clusters, for example, that regions 1 and 4 in the top experiment correspond to 2 and 3 in the second, then we can split the sequences between this points and align all the resulting subsequences, as shown in Fig. 7b. Now if we only pay attention to the order of occurrence of the computations, all those that appear in the same column are equivalent with respect to their chronological order.

In order to decide which are the points to split the sequences, this heuristic uses the matchings discovered so far by the previous heuristics to establish pivots in both sequences, and align the subsequences with respect to these points of reference to discover new matchings.

3.2   Combining Tracking Heuristics

Build upon the combination of these five heuristics, the tracking algorithm proceeds as follows to determine a global matching between all clusters. Every heuristic is applied separately and reports one or more correlation matrices representing relations between objects. Depending on the heuristic, what these matrices express is different. Figure 4 shows the correlations computed by the *density* heuristic for experiments WRF-128 (A) and WRF-256 (B). In this case, it indicates the percentage of computations that conform object A_i for which object B_j is closer. As you can see, there are cases where one object is close enough to two others or more, so it is not immediate to determine the appropriate correspondences when the objects are moving arbitrarily around the performance space. For the *SPMDiness* heuristic one correlation matrix per frame is built, each expressing the probability of two different computations to be executed at the same time by different processes within the same experiment. The *call stack* heuristic calculates the percentage of computations that are part of object A_i whose call stack references point to the same source code than those of object B_j. The *density* heuristic represents groups of clusters that have the same aggregate density. Lastly, the *chronological sequence* heuristic reflects the percentage of times where computations A_i and B_j happen in the same order of occurrence. In all cases, non-zero cells evince that a given pair of objects are the same with a certain probability, according to that heuristic. Occurrences with a very small probability (5 % by default) are neglected as outliers.

   Since every heuristic considers different properties of the objects, there might be contradictions on the correspondences found, and the results have to be combined to complement the correspondences that a given heuristic might fail to discern. To this end, a combination algorithm extracts from each correlation matrix a set of rules in the form $A_i \equiv B_j$ expressing which objects between two frames are equivalent, and reduces the rules applying a series of union and intersection operations. The union operation computes a logical *OR* and can be seen as complementing the results of different heuristics (e.g. one heuristic finds that $A_1 \equiv B_1$, and another finds that $A_1 \equiv B_2$, so we add up the results and consider that $A_1 \equiv B_1 \cup B_2$). In this case, an equivalence between two objects is kept always that at least one heuristic confirms it. The intersection operation computes the logical *AND*, and can be seen as the agreement between heuristics (e.g. in the previous example, there is no valid correspondence for A_1 because the heuristics did not agree). In this case, an equivalence between two objects is kept only if all the heuristics find that same correspondence, or discarded otherwise.

   The first rules to take into account are always those found by *distance*, because the information required to compute the distances between objects is always present in the frames. Then the resulting rules are united with those found by *SPMDiness*. For example, if the first finds that the nearest object for A_5 is B_5, and the latter finds that B_5 and B_{13} always happen simultaneously, all objects merge into a more general relation $A_5 \equiv B_5 \cup B_{13}$. The *call stack* and *density* rules are then intersected to prune incorrect relations that may appear due to mapping errors in the former heuristics.

For example, all related clusters must share the same references to the source code, so we discard those not having any in common.

We search for correspondences between objects reciprocally, this is to say, comparing frame *A* with *B* and vice versa, extracting a final set of rules that correlate the objects between both frames. When the information available leads the heuristics to not be able to clearly distinguish one region from another, the regions in doubt are grouped together, resulting in wide relations of multiple objects. The *chronological sequence* heuristic is finally used to refine the results, splitting wide relations into more specific ones.

The analysis is repeated for every pair of consecutive frames, obtaining in the end *k tracked regions*, relations of objects that are equivalent along the whole sequence of images. Additionally, the tool generates plots describing the evolution of each *tracked region*. Next section gives an overview of the results of the tracking algorithm.

3.3   Tracking Results

In this section we present the results of the tracking algorithm, following on from the WRF example used to guide the explanation of the technique through the former sections. For the two configurations presented, runs with 128 and 256 tasks, we will conduct a brief scalability study to explain how the tracking results yield practical insights that help in understanding and improving the code.

First, the tool reconstructs the input images for the tracking algorithm with all objects identifiers renamed, so that all equivalent regions keep the same numbering and color. The whole sequence of images can be displayed in a simple animation, or in a single plot showing the trajectory that every different object follows, so that is very easy to identify variations in the performance space, as shown in Fig. 8 (in logarithmic scale for better readability, refer to Fig. 2 for the real scales).

Here we can observe two main trends: clusters whose shape hardly varies between experiments (e.g. Regions 1 to 3), and those that become more distorted when the scale increases (e.g. Regions 4, 5 and 7). Focusing on the latter which are most affected by the scale, the developers made an effort to balance the amount of work, as they appear as flat clusters with low variation in the instructions axis. However, they present large IPC variability that increases at higher scale. In the 256-tasks case Regions 4, 5 and 7, that cover altogether the 30 % of the total time, split into new zones of imbalance on their left with lower performance. Clusters becoming more disperse indicate an increasing problem of time imbalance.

Amongst the regions that do not deteriorate due to the scale increase, Region 2 stands out for covering all alone the 15 % of the execution time, and exhibiting an elongated cluster in the Y-axis that reflects large instructions imbalance, within a dynamic range that doubles from 1.5e9 for the 128-tasks case (top), and 8e8 for the 256-tasks case (bottom). Despite the IPC variability partially compensates the

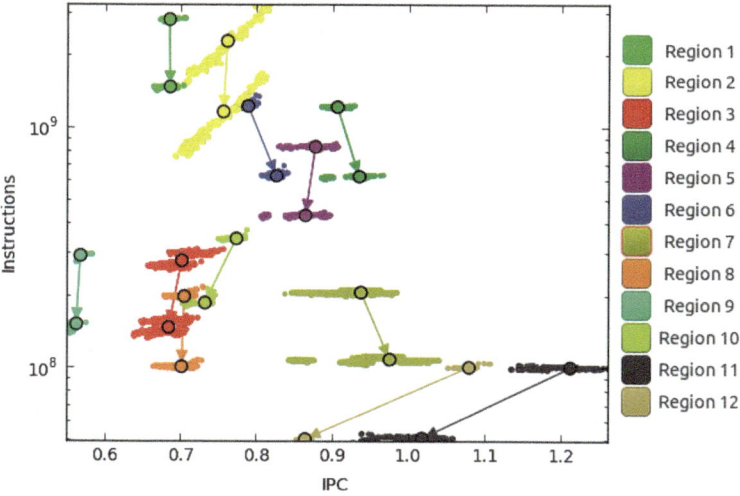

Fig. 8 Trajectories of clusters from WRF-128 to WRF-256

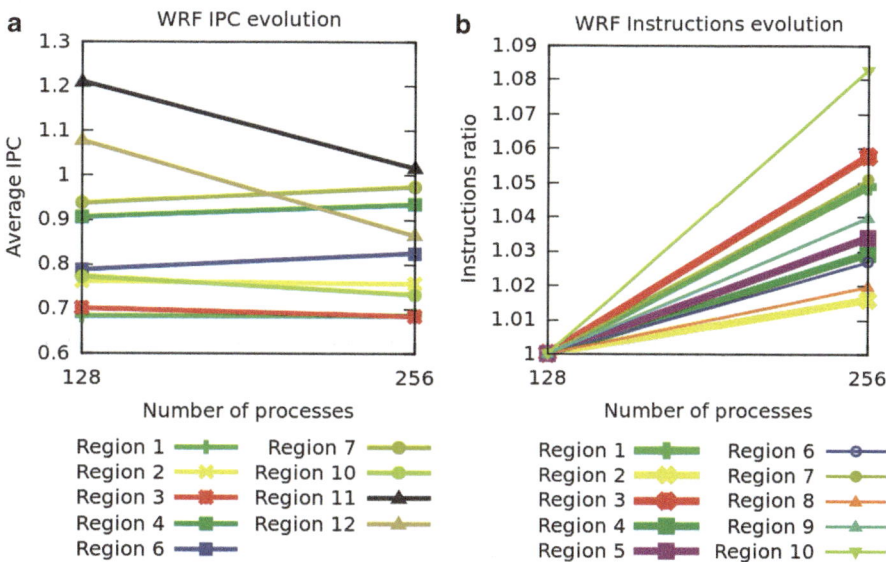

Fig. 9 Performance trends for WRF code regions. (**a**) IPC evolution. (**b**) Instructions evolution

instructions imbalance and the performance is maintained at scale, this region was already inefficient from the start.

In addition, the tool presents the evolution of every computing region from the first scenario to the last, with respect to the metrics selected to generate the images. Figure 9a shows a trend chart displaying the evolution in IPC for the 128 and 256-tasks runs of WRF. For better readability, only the most significant regions and those with higher IPC variations (above 3 %) are depicted. While there is a slight improvement for regions 4, 6 and 7 under 4 %, regions 10 to 12 present a sharp decline up to 20 %. Regions 1 to 3 remain constant, yet is important to remark that being the most important computations covering 50 % of the total time, these are also the ones achieving lower IPC around 0.70. Figure 9b shows the evolution in the number of instructions for the regions that execute the most, as the percentage over the 128-tasks base case. When the number of cores increases, so does the total number of instructions, revealing code replication below 8 % in all regions of the program, which is reasonable but warns us about an increasingly detrimental effect at higher scales, in particular for regions 3 and 10.

For a production class application with a long-term development, a brief analysis of the clusters trajectories and the metrics trends has quickly diagnosed several performance weaknesses and potential problems at higher scales. In general, the information presented allows to perform parametric studies on the influence of different configurations, as well as to study the evolution of a single experiment over time, enabling an intuitive analysis that gets straight to the points of interest and their major causes of inefficiency. Having call stack references associated to every cluster, it is possible to connect the observed performance artifacts to specific points in the code and extract useful recommendations on which way to direct the optimization process.

4 Cases of Study

The aim of this section is to demonstrate the added value of using tracking, where the importance lays on understanding how and why the performance of the application changes along multiple experiments. We want to highlight the versatility of the technique for a variety of parametric studies, tossing ideas about the kind of cases of study that could be interesting for the analyst. To this end, we have selected configurations that would produce unpredictable sets of clusters and arbitrary displacements to prove the algorithm working under stress. Moreover, we present a real-case study to show that this technique can be useful to provide valuable insights to the users and successfully lead to improvements in their codes.

Therefore, a variety of proxy and production codes from different fields such as astrophysics, molecular dynamics and meteorology; were run in MareNostrum II, MareNostrum III and MinoTauro [2]. MareNostrum II is a cluster of 2,560 nodes, each containing 2 IBM PowerPC 970MP 2-Core at 2.3 GHz with 8 GB of RAM. MareNostrum III comprises 3,028 nodes, each containing 2 Intel SandyBridge-EP E5-2670 8-Core at 2.6 GHz with 32 GB of RAM. MinoTauro comprises 126 nodes, each containing 2 Intel Xeon E5649 6-Core at 2.53 GHz with 24 GB of RAM.

Table 4 illustrates the ability of the algorithm to identify and keep track of the different computing regions in 11 studies. The objects detected are automatically reduced to the ones considered more relevant, those that represent a high percentage of the total application time, usually above 5–10 %. *Coverage* is calculated as the

Table 4 Summary of experiments

Application	Input images	Tracked regions	Coverage
Gadget	2	8	88 %
QuantumE	2	6	66 %
WRF	2	12	100 %
Gromacs	3	5	100 %
CGPOP	4	2	66 %
NAS BT	4	6	100 %
OpenMX	7	7	100 %
Hydro	8	3	100 %
MR-Genesis	12	2	100 %
NAS FT	15	2	100 %
Gromacs	20	4	80 %

percentage of objects tracked with respect to the maximum number of identifiable objects in the input images. 100 % in *coverage* denotes that the algorithm has been able to find unambiguous correspondences between all the objects. Values below the optimal reflect that there were nearby objects in the input images that the tracking heuristics could not distinguish as separate individuals with the information available, grouping them as a single entity. On average, the algorithm successfully discriminates 90 % of the objects. The following sections present seven case studies in more detail.

4.1 Studying the Scalability of the Computing Regions

The objective of this experiment is to conduct a real-case study of the scalability of the computing regions of an application. The selected code is OpenMX [19], a software package designed for the realization of large-scale ab initio calculations. To this end, we run OpenMX v3.6p1 in MareNostrum III increasing the number of MPI tasks from 64 to 512 using a single OpenMP thread per task.

As we are running a strong-scale test (fixed-size problem on a varying number of processors), the application would ideally see the execution time reduced inversely proportional to the number of processors used. However, multiplying by 8 the number of tasks, the speedup achieved in a single time-step is lower than 2. In terms of work executed, the total number of instructions should have got evenly distributed amongst all processes, and thus remain constant when the scale increases. Withal, Fig. 10a shows the total number of instructions increasing by 100 % from the 64 to the 512-tasks case, which is far from the ideal scaling and too significant to be due to a problem of code replication. Applying tracking, we can now break-down this aggregate for the whole program and study the evolution of the relevant code regions per separate, to understand which parts prevent the application from scaling better.

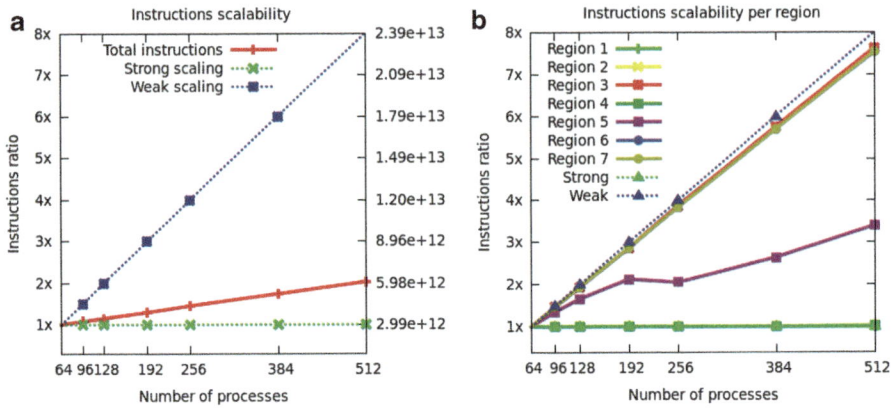

Fig. 10 Scalability of OpenMX. (**a**) Application scalability. (**b**) Computations scalability

The input to the tracking algorithm is the collection of images that depict the performance of each individual experiment. Unlike in other experiments where the images are two-dimensional (Instructions and IPC), in this case we used the metric *L1 data cache misses* as a third dimension to cluster the data, which results in a more precise characterization of the relevant computational behaviors. Figure 11a shows the result of the tracking algorithm applied to the sequence of experiments from 64 to 512 tasks (only 6 out of 7 depicted due to space constraints, and plotted in 2D in the Instructions and IPC axes for clarity).

A quick glance at the evolution of the main behaviors reveals two main issues: First, most regions progress vertically downwards the Y-axis (instructions decrease), as one would expect for a strong-scaling case. Figure 11b shows the trajectories that follow the different regions from one experiment to the next, represented by their centroids. It is easy to see that regions 3, 6 and 7 do not move, meaning that they perform constant work despite the scale, as if they were ran in a weak-scaling mode.

Figure 10b shows the ratio of surplus work executed per region with respect to the ideal case where all regions scaled perfectly. In the 512-tasks case, regions 3, 6 and 7 which should have seen reduced their work by a factor of 8, actually execute 7.5 times more work than the expected. With this progression, these three regions that represented altogether the 20 % of the iteration time in the 64-tasks case, now dominate the iteration representing the 65 % of the total time, and have become the main bottlenecks to the computation scalability. Namely, these correspond to the computing phases starting at lines 289, 589 and 129 of routine Set_XC_Grid. Here, the programmer has put effort to use shared memory programming, but has not taken advantage of distributing the workload amongst processes. Likely, the developers considered more efficient to replicate this code to avoid the cost of communications, which may be worthwhile at small scales, but the increasing costs do not pay off at larger scales. These observations were reported to the developers, suggesting to study the feasibility of partitioning the work so as to fully exploit the distributed resources.

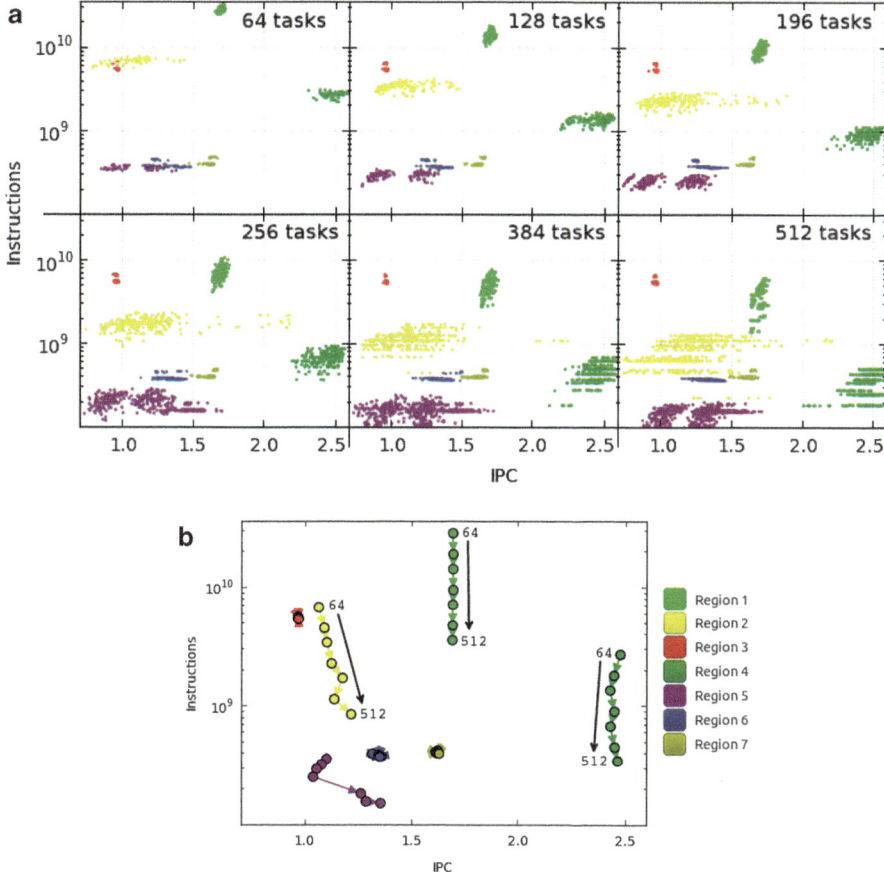

Fig. 11 Tracking results for OpenMX. (**a**) Sequence of output images from the tracking algorithm. (**b**) Trajectories of clusters from 64 to 512-tasks runs

The second important observation is that most behaviors grow more and more disperse. In particular, it is the regions that scale better the ones that present more variability, namely 1, 2 and 4. The *parallel efficiency* [5] of these regions decreases from 0.80 in the 64-tasks case to 0.60 in the 512-tasks case, meaning that the 40 % of resources are wasted due to time imbalances, where some processes have to wait for others to finish their work, and such imbalance gets absorbed in subsequent synchronizations. These correspond to the computing phases starting at lines 732 of Krylov_Col, 256 of Set_Hamiltonian, and 288 of Set_Density_Grid. In this case, a second precise recommendation could be made to the user to study the load-balancing characteristics of these particular regions.

As a final remark, the detected hazards could have been inferred just from the first three frames in the sequence, and so our technique can be used with few cores to anticipate problems at higher scales, saving on time and resources.

4.2 Studying the Impact of Multi-core Sharing

MR-Genesis [17] employs a finite volume approach in order to evolve the Relativistic Euler equations combined with a Constrained Transport scheme to account for the divergence free evolution of the dynamically included magnetic field. MR-Genesis was run in MinoTauro using 12 processes, changing the maximum number of processes allowed per node from 1 to 12. Being 12 the number of available cores per node in MinoTauro, the configuration for the first experiment corresponds to 12 different nodes running a single process each, and a single full node for the last experiment, with all the intermediate cases also tested. The objective is to study the effect of memory bandwidth and caches contention on the application performance when sharing resources.

Figure 12a shows the result of the tracking algorithm applied to the sequence of experiments from 1 to 12 processes per node, which reveals two main computing phases with analogous behavior. Since it is only the physical mapping of processes what changes, the total number of instructions executed remains constant in all trials. However, as nodes get more populated, the achieved performance of the application decreases. Up to the 66 % of the node occupation (8 tasks per node) the IPC presents a slight reduction under 1.5 % from one experiment to the next, but starts presenting sharper drops beyond this point, with an 8.5 % loss when an additional process is collocated in the node. Overall, the achieved IPC degrades a total of 17.5 % when the node is full.

Figure 12b correlates all performance metrics for Region 1. The Y-axis reflects the percentage of variation of each metric with respect to its maximum value for all trials. The number of L2 cache misses grows inversely to the IPC degradation rate, and the TLB misses also increase as the node gets more populated.

In this case, a fair trade-off between maximum utilization of the resources and the application performance is met at two-thirds of the node occupation.

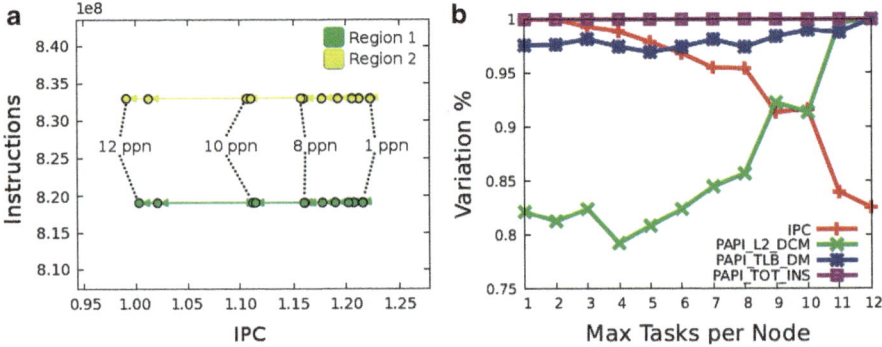

Fig. 12 Tracking results for MR-Genesis. (**a**) Clusters trajectories mapping from 1 to 12 processes per node. (**b**) Region 1 evolution

4.3 Studying the Impact of the Program Block Size

HYDRO [15] is a proxy benchmark that solves a large scale structure and galaxy formation problem using a rectangular 2D space domain split in blocks. HYDRO was run in MinoTauro, and the sequence of images in this case is built doubling the block size from 8 to 1,024 KB. The objective of this experiment is to determine which is the best setting for a particular parameter of the program to minimize the execution time.

Figure 13a shows the evolution of the three main computing phases of the application, which actually refer to the same source code region with tri-modal behavior. The trajectories reflect the number of instructions initially decreasing for all three regions with drops from 1 to 3 % up to a block size of 32 (movement downwards the Y-axis), and keeps steady beyond this point. IPC also decreases with a total deviation of 5 % for Region 1, and 10 % for Regions 2 and 3, all presenting a sharp dip when the block size increases from 64 to 128 (movement leftwards the X-axis). At this point, the number of L1 data cache misses rockets 40 % more, as shown in Fig. 13b.

Using small block sizes the application gets more blocks to compute, which entails executing more control instructions. Since the blocks are bi-dimensional and store 8-bytes elements, when the block size is set to 64 the limit of the L1 cache is reached, which is 32 KB. With bigger sizes, the block does not fit in the cache, and so the miss rate increases to the detriment of IPC.

Correlating the evolution of all metrics, the point where highest performance and lowest workload and cache misses converge is at a block size of 16, which results in the fastest execution of all proposed setups, so this one would be the most recommendable to minimize the response time.

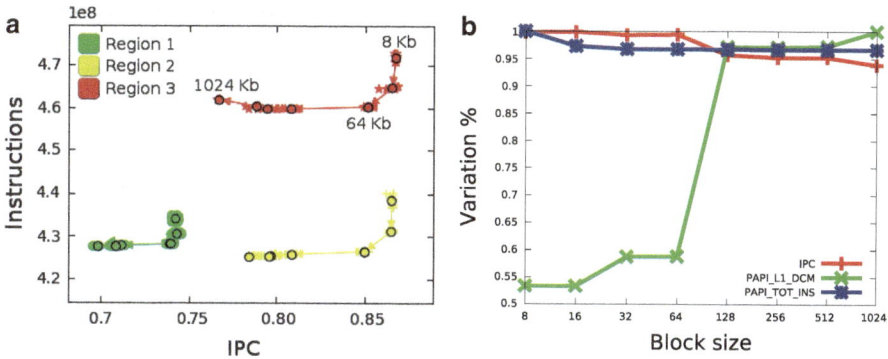

Fig. 13 Tracking results for Hydro. (**a**) Clusters trajectories doubling the block size from 8 to 1,024 KB. (**b**) Region 1 evolution

4.4   Studying the Impact of the Problem Input Size

The NAS Parallel Benchmarks [18] are a small set of programs designed to assess the performance of parallel supercomputers. In this experiment we evaluate version 2.3 of the BT solver with increasing problem sizes. Problem sizes are predefined and indicated as different classes, where Class W corresponds to a small workstation problem size, and A, B and C correspond to standard test problems with a 4X size increase going from one class to the next. For all classes, BT was run in MareNostrum II with 16 processes.

Figure 14 shows the trajectories of the clusters through classes W to C. The starting experiment corresponds to Class W, which can be located at the bottom part of the plot. Class W presents large variability in IPC, which is depicted with the elongated clusters in the X-axis. As the experiments move forward, all clusters move to the top-left part of the plot. This transition shows a large dynamic increase of two orders of magnitude in the number of instructions from Class W to Class C. Also, clusters become more compact, indicating a reduction in the IPC variability except for Region 2, which corresponds to the Gaussian elimination performed in routines $[x|y|z]\_solve\_cell$.

In contrast, the achieved performance in all code regions degrades as the size of the problem increases. Figure 15a shows there are two decreasing trends for the IPC. For regions 1, 2, 4 and 5, a sharp loss ranging from 40 to 65 % happens as soon as we move from Class W to A and then stabilizes, while for regions 3 and 6 the IPC keeps decreasing and does not stabilize until Class B. Correlating the evolution of all available metrics, we can see that this IPC degradation can be explained due to an increase in the L2 data cache misses, as shown in Fig. 15b.

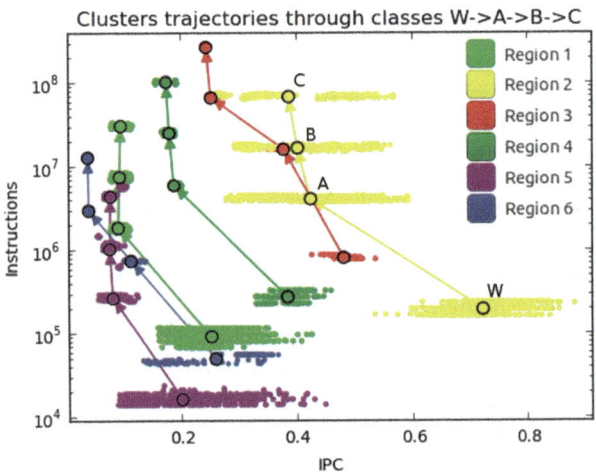

Fig. 14 Trajectories of clusters through BT experiments

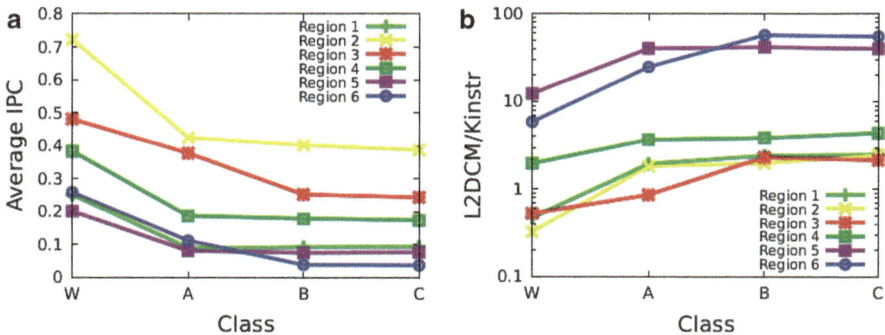

Fig. 15 Performance trends for NAS BT code regions. (**a**) Evolution of IPC. (**b**) Evolution of L2 cache misses

4.5   Studying the Impact of Different Hardware and Compilers

In this experiment we are going to stress the performance variations in the application changing the machine where it is executed and also changing from a generic to an architecture specific compiler. This test shows that even in very different scenarios that may result in large performance variations, the tracking algorithm is able to follow the evolution of the clusters.

CGPOP [27] is a proxy application of the Parallel Ocean Program [14]. POP simulates the global climate model and is a component of the Community Earth System Model. CGPOP was run with 128 processors both in MareNostrum II and MinoTauro, and compiled with GNU Fortran 4.1.2 (gfortran) and IBM XL Fortran 12.1 (xlf) in MareNostrum, and GNU Fortran 4.4.4 and Intel Fortran 12.0.4 (ifort) in MinoTauro. In all cases, the application was compiled with an aggressive optimization flag ($-O3$) and debug ($-g$).

Figure 16 shows the trajectories that follow the two main computing behaviors with respect to the number of instructions, which are subdivided into several regions due to differences in the achieved IPC. In MareNostrum, when the application is compiled with xlf (see Fig. 16b) all computations see the number of instructions significantly reduced (36 and 33 %, respectively) compared to using gfortran (see Fig. 16a), but the IPC degrades practically in the same proportion and the overall execution time remains almost constant. The situation in MinoTauro is very similar (see Fig. 16c, d), with an overall improvement in terms of less instructions executed and higher IPC achieved, yet the same degradation effect when changing compilers can be easily identified.

Changing the platform also alters the behavior of code, as can be seen for Region 2 in MareNostrum which splits into Regions 2 and 3 in MinoTauro, no matter the compiler used. They all refer to the same point in the code, but it now presents two distinct behaviors. The tracking algorithm automatically identifies and groups together those regions that are equivalent despite the performance variations, as

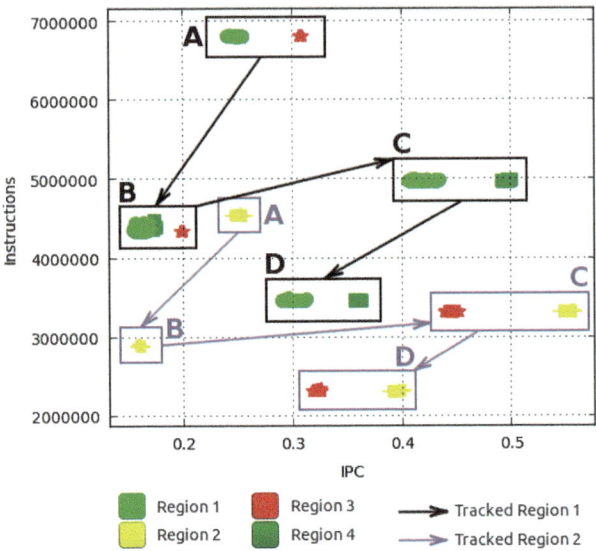

Fig. 16 Trajectories of clusters through CGPOP experiments: (*A*) MareNostrum GNU (*B*) MareNostrum IBM XL (*C*) MinoTauro GNU (*D*) MinoTauro Intel

Table 5 CGPOP performance results

		MareNostrum II		MinoTauro	
		gfortran	xlf	gfortran	ifort
Region 1	IPC	0.25	0.16	0.42	0.30
	Instructions	6.8 M	4.3 M	5 M	3.5 M
	Duration	12.09 s	12.11 s	4.82 s	4.68 s
Region 2	IPC	0.25	0.16	0.50	0.36
	Instructions	4.5 M	3 M	3.3 M	2.3 M
	Duration	2.13 s	2.14 s	0.71 s	0.69 s

illustrated by the bounding boxes, and then numerically calculates their evolution along experiments. Table 5 summarizes the averages for IPC and instructions for both tracked regions, and their elapsed execution time.

In this case, the specialized compilers xlf and ifort attain a reduction of 36 and 30 % of the number of instructions with respect to gfortran in both machines, but at the expense of an average IPC loss of 36 % in MareNostrum and 28 % in MinoTauro. Likely, they reduced index arithmetic but the performance did not change much because the computation is still memory bound. The integer instructions saved were likely traded for idle issue slots while waiting for the memory hierarchy, leading to negligible variations in the execution times lower than ±0.03 %.

4.6   Studying the Effect of Optimal and Non-optimal Grid Geometries

HACC (Hardware/Hybrid Accelerated Cosmology Code) is a framework that melds particle and grid methods to satisfy the requirements of cosmological surveys, exploiting hybrid and accelerator-based architectures with millions of cores, including CPU/GPU, multi/many-core, and Blue Gene systems. HACC is designed to scale weakly by dividing the working data set in cubes. In this experiment we stressed the application setting different geometries other than a perfect cube, in order to see how much is the performance affected. The program was run in MareNostrum III, doubling the number of tasks from 16 to 1,024 tasks, as well as the size of the problem, with 1 single MPI task per node (so neither multi-core nor memory caches sharing), using the Intel MPI message passing library, and without support for threads.

Figure 17 shows the trajectories of the main computing regions of HACC. Here we can observe zigzag movements back and forth: as we increase the number of tasks (and so the size of the problem in proportion), all regions move upward (more instructions executed) and rightward (more IPC achieved). However, when the number of tasks is cubic (i.e. 64 and 512 tasks), the regions move back in the opposite direction (down and left; meaning less instructions executed and less IPC achieved). Figures 18 and 19 show this effect more clearly. Figure 18 shows the

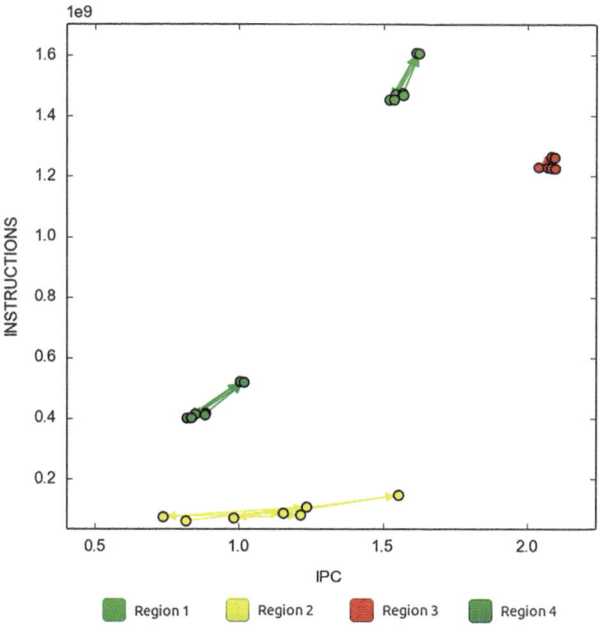

Fig. 17 Trajectories of clusters through HACC weak scale experiments

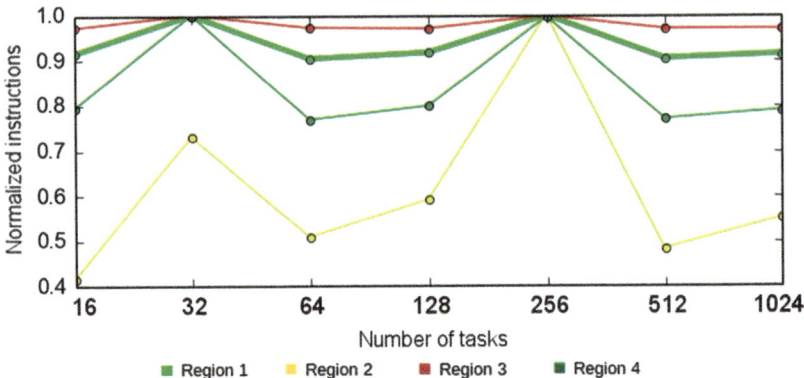

Fig. 18 Average instructions executed per computing region in HACC

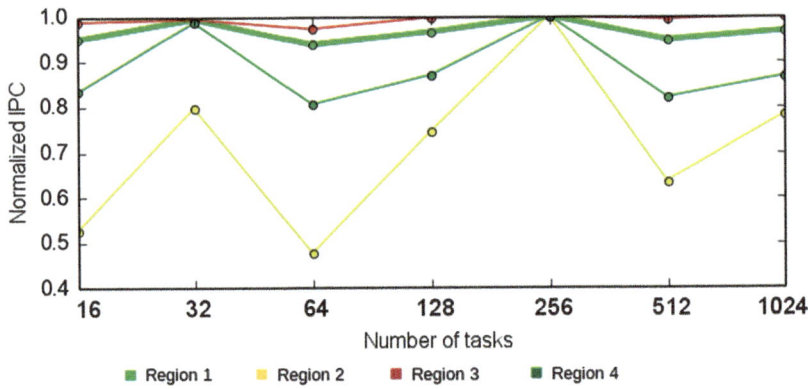

Fig. 19 Average IPC achieved per computing region in HACC

amount of instructions executed per region across experiments. The lower workload is found at experiments 3 and 6 (the cubic cases with 64 and 512 tasks). Correlating with Fig. 19, these two experiments are also the ones achieving lower IPC.

The differences in the number of instructions can be explained due to the work distribution scheme: when the number of tasks is not cubic there is extra work to distribute among the available tasks. Although the IPC achieved also becomes higher, the increase in performance does not compensate for the increase of work, and the computation time becomes higher in the uneven cases. This can be seen in Fig. 20, that compares the computation times for the main computing region of the program (Region 1) in the cubic runs (64 and 512 tasks), and an uneven intermediate case (256 tasks). In these histograms, the rows represent processes and the columns are bins of computation durations increasing from left to right, and the colors represent the time spent on a particular bin, ranging from green (low time) to blue (high time). In the cubic cases (see Figs. 20a, c), the computing times are very similar in the range of 310 to 323 ms, but in the 256 tasks case (see Fig. 20b), all

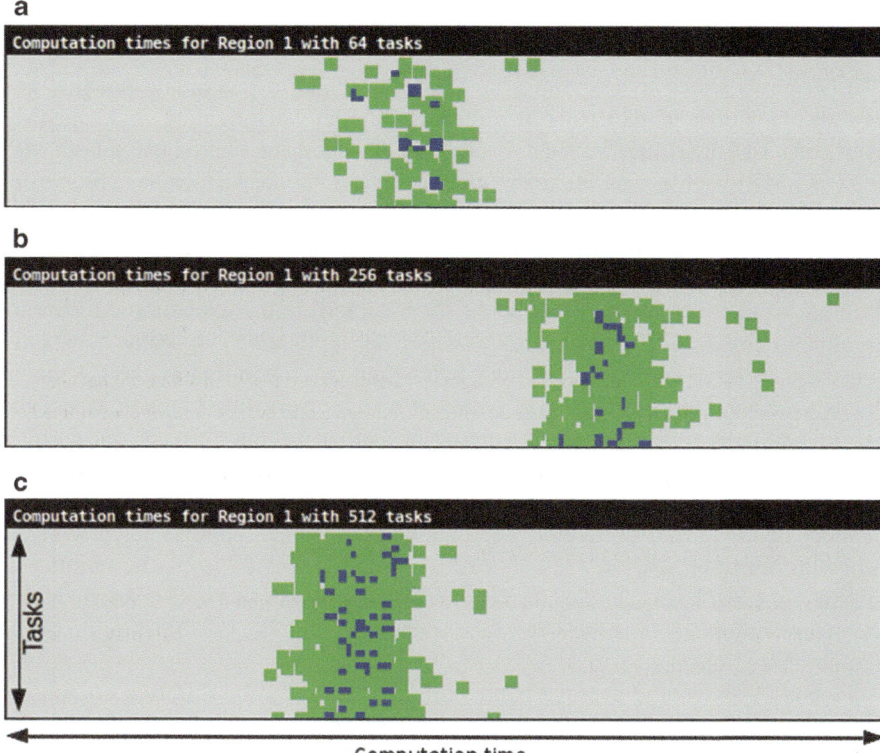

Fig. 20 Histograms of durations for computing region 1 in HACC. (**a**) 64 tasks. (**b**) 256 tasks. (**c**) 512 tasks

computations are shifted to the right, having increased their times ranging from 323 to 343 ms.

Even though the overall performance of the computations is better in the cubic cases, we can also observe that the time to solution degrades as we increase the scale. Comparing the two cubic cases with 64 and 512 tasks, we see that the percentage of time spent in computations decreases from 60 to 45 %, and so the communications start to dominate the execution time. Analyzing the trace, the time spent in MPI_Wait calls increases from 30 to 50 % because of the serializations in the program caused by a pipelined communication pattern, where some processes can not progress until they have received messages they are waiting on. One recommendation that could be given in this case to improve the scalability of the program is to change the communication pattern so as to overlap communications with computations, reducing the serializations.

4.7 Studying the Effects of Memory Congestion

LULESH is a shock hydro mini-app. While designed to test many machine and
hardware features in particular it stresses compiler vectorization and OpenMP
overheads. LULESH performs a hydrodynamics stencil calculation using both MPI
and OpenMP to achieve parallelism. In general the compute performance properties
of LULESH are more interesting than messaging as on a typical modern machine
only about 10 % of the runtime is spent in communication.

In this experiment we measure how sensitive are the computations of this
program to memory congestion. To do so, we interfere the execution collocating
gremlin processes [6] in the same nodes where the application is running, constantly
consuming a large amount of memory bandwidth by contaminating the L3 shared
memory cache. One gremlin is activated at a time every few seconds across the
execution. So the application starts running clean, then it is interfered by one gremlin
after a while, then by a second gremlin and so on until a maximum of six gremlins
per socket and node are interfering. Every single gremlin consumes an average
memory bandwidth of 600 Mb/s. The bandwidth has been approximated with the
following formula: $BW = (L3_{miss} * L_{size})/T_{burst}$; where $L3_{miss}$ is the number of
misses in the last level cache, and so is the number of times that data is fetched from
the main memory; L_{size} is the size of the cache line, and T_{burst} is the duration of the
computing region.

Figure 21 shows the average bandwidth consumed by the different computing
regions of LULESH with increasing number of gremlins. As the reader can see,
each new gremlin increases the average bandwidth consumed by the application

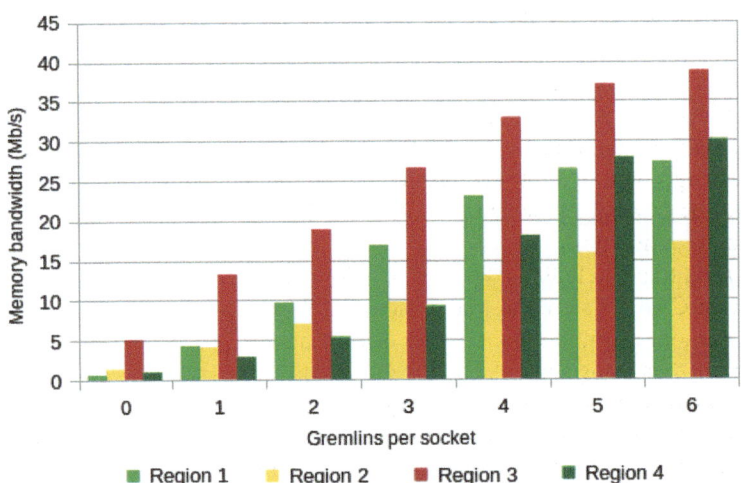

Fig. 21 Memory bandwidth consumed by the main computing regions of LULESH with increasing levels of interference

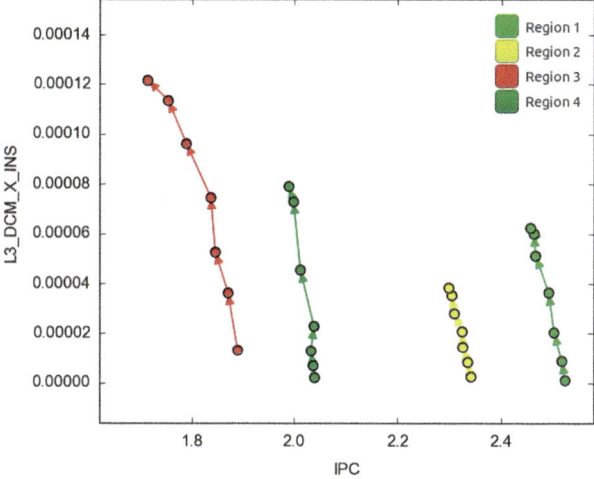

Fig. 22 Trajectories of clusters through LULESH experiments (from *bottom* to *top*, 0–6 gremlins per socket)

due to the extra memory accesses needed to cope with the higher number of cache misses.

Figure 22 shows the trajectories of the main four computing regions of LULESH with respect to the IPC achieved (X-axis) and the number of L3 misses (Y-axis). All regions move upwards (meaning that the number of L3 cache misses increase with the number of interfering gremlins), and also move leftwards (meaning that all regions loose performance with higher levels of interference). Region 3 (red) is the one that moves further in both axes, which means that is the computation most affected by the interferences, and corresponds to the computations performed at the CommMonoQ routine. This routine copies data from several structures with a non-consecutive stride, making more misses because the application is not taking advantage of the temporal data locality.

5   Related Work

Our work draws inspiration from a motion detection algorithm of moving biological objects that are similar but non-homogeneous [20]. They apply multi-feature contour segmentation and flux tensors for identifying the boundaries of biological objects and the detection of deformable motion and complex behaviors (e.g. cell crawling or division) along a time-lapse collection of images.

In a broader sense, object tracking is applied in the context of applications that require to associate target objects in consecutive frames to detect how they move around the scene. Practical applications include: automated surveillance, gesture recognition, traffic monitoring or path planning and obstacle avoidance. Yilmaz et al. [30] presents an extensive review of the state-of-the-art of tracking methods, and discusses related issues including the use of appropriate image features, motion models and object recognition.

ETRUSCA [22] is a post-mortem performance tool that includes a jitter reduction analysis that attempted to relate the clusters found in one time interval with the clusters found in the next interval. Selecting a representative process in each interval, they would minimize the data captured. Our approach does not look for representative processes, but representative behaviors for all computing phases within all processors, and track how they change not only across time intervals, but also across experiments with different configurations.

Multi-experimental analysis has been approached by several performance analysis tools. SCALASCA [26] includes a tool called performance algebra that can be used to merge, subtract, and average the data from different experiments and view the results as a single derived experiment. PerfExplorer [12] supports data mining analyses on multi-experiment parallel performance profiles. Its capabilities include general statistical analysis of performance data, dimension reduction, clustering and correlation of performance data, and multi-experiment data query and management. TAU [25] incorporates the concept of phase profiling for the study of the evolution within a single experiment. This is an approach to profiling that measures performance relative to a phase of execution, having its entry and exit marked by the user. HPCToolkit [16] merges profile data from multiple performance experiments into a database file and perform various statistical and comparative analyses.

While they compute averages for predefined metrics and fixed phases such as functions, iterations or sections marked beforehand, we report arbitrary metrics at the level of computing regions. By doing so, we abstract the structure of the application to the behavior of its computing phases, taking into account the performance measurements of every single computation rather than profiled averages that may hinder their actual behavior.

The fundamental difference that distinguishes our approach from the previous ones is that we do not merely report the outcome of different experiments together. We automatically determine the regions of interest and track their evolution along multiple executions. To this end, we translate performance data from different execution scenarios into a sequence of images, detect structure in each image and automatically correlate them.

6 Closing Remarks

In this chapter we have demonstrated that it is possible to draw an analogy between tracking techniques applied to the automatic detection of an object's motility, and the performance analysis of a parallel application's evolution along multiple execution scenarios. This approach mimics the common phase structure of a tracking algorithm, including the generation of a sequence of images, object recognition within each frame and motion analysis across scenes.

Different scenarios are represented as a sequence of performance images that expresses the evolution of the application either along different experiments with changing configurations, or along time intervals within the same experiment. Computing regions of the application are represented as objects in these images, described by how they behave in terms of selected performance metrics. Then, we find a correspondence between objects along the whole sequence of images, keeping track of their possible motions and structural changes due to performance variations. To this end, we use a variety of heuristics that take into account different characteristics of the computing regions: the distances in the performance space, the SPMDiness of the application, the code region they refer to, the clusters densities and the chronological sequence. Combining their use, we are able to automatically identify the global evolution of the main computational behaviors and illustrate their performance trends.

Our technique offers a different viewpoint to the task of analysis that is more agnostic of the syntax of the code, but brings into focus the main performance characteristics of the program and the nature of their inefficiencies, enabling the identification of the most appropriate solution for the bottlenecks observed. Then, these observations can be correlated with the source code, to know which sections exhibit a given behavior. There are two remarkable benefits to this approach. First, the same solution can be applied to multiple code sections that present the same deficiency, without having to reappraise the same problems repeatedly. Second, we are able to detect multi-modal behavior and variations along time and processors, two important effects often masked by profiling tools. In this way, a single code section undergoing performance variabilities will be expressed as divergent behaviors that can be studied separately, revealing more room for improvement.

All in all, this work presents a versatile tool applicable in very varied scenarios, enabling the analyst to study the impact of virtually any configuration on the application performance without prior knowledge of the program; compare and correlate performance data between experiments; determine the best setup to meet specific performance requirements; and ultimately helps to gain better understanding of the application behavior, much beyond what can be learned from a single experiment.

This work opens up interesting lines of future research. On one hand, predictive models could be built next that would enable us to foresee the performance of experiments beyond the sample space. On the other hand, further on-line integration could be developed, in order to analyze the evolution over time of adaptive applications automatically.

References

1. Browne, S., Dongarra, J., Garner, N., Ho, G., Mucci, P.: A portable programming interface for performance evaluation on modern processors. Int. J. High Perform. Comput. Appl. **14**, 189–204 (2000)
2. BSC Facilities. http://www.bsc.es/marenostrum-support-services
3. BSC Tools. http://www.bsc.es/paraver
4. Buck, B., Hollingsworth, J.K.: An API for runtime code patching. Int. J. High Perform. Comput. Appl. **14**, 317–329 (2000)
5. Casas, M., Badia, R., Labarta, J.: Automatic analysis of speedup of MPI applications. In: Proceedings of the 22nd Annual International Conference on Supercomputing (ICS'08), Island of Kos, pp. 349–358. ACM, New York (2008)
6. Casas, M., Bronevetsky, G.: Active measurement of memory resource consumption. In: Proceedings of the 2014 IEEE 28th International Parallel and Distributed Processing Symposium (IPDPS'14), Phoenix, pp. 995–1004. IEEE Computer Society, Washington, DC (2014)
7. Geimer, M., Saviankou, P., Strube, A., Szebenyi, Z., Wolf, F., Wylie, B.J.N.: Further improving the scalability of the SCALASCA toolset. In: Proceedings of the 10th International Conference on Applied Parallel and Scientific Computing (PARA'10), Reykjavik, vol. 2, pp. 463–473. Springer, Berlin/Heidelberg (2012)
8. GNU Binutils. http://www.gnu.org/software/binutils
9. González, J., et al.: Automatic detection of parallel applications computation phases. In: IPDPS: 23rd IEEE International Parallel and Distributed Processing Symposium, Sao Paulo (2009)
10. González, J., et al.: Automatic evaluation of the computation structure of parallel applications. In: PDCAT: Proceedings of the 2009 International Conference on Parallel and Distributed Computing, Applications and Technologies, Higashi Hiroshima, pp. 138–145 (2009)
11. González, J., et al.: Performance data extrapolation in parallel codes. In: ICPADS: 16th IEEE International Conference on Parallel and Distributed Systems, Shanghai, pp. 155–163 (2010)
12. Huck, K.A., Malony, A.D.: PerfExplorer: a performance data mining framework for large-scale parallel computing. In: Proceedings of the Conference on Supercomputing, New York, p. 41 (2005)
13. Ibarra, O.H., Kim, C.E.: Fast approximation algorithms for the knapsack and sum of subset problems. J. ACM **22**(4), 463–468 (1975)
14. Jones, P.: Parallel Ocean Program (POP) user guide. Technical report, Los Alamos National Laboratory, March 2003
15. Lavalléea, P.-F., et al.: HYDRO. http://www.prace-ri.eu
16. Mellor-Crummey, J.: HPCToolkit: multi-platform tools for profile-based performance analysis. In: APART, Nov 2003
17. Mimica, P., Giannios, D., Aloy, M.A.: Deceleration of arbitrarily magnetized GRB Ejecta: the complete evolution. Technical report arXiv:0810.2961, Oct 2008. Comments: 13 pages, 10 figures, revised version sent to the referee (first version submitted on 6th of August)
18. NAS Parallel Benchmarks. http://www.nas.nasa.gov/Software/NPB
19. Open source package for Material eXplorer. http://www.openmx-square.org
20. Palaniappan, K., et al.: Moving object segmentation using the flux tensor for biological video microscopy. In: PCM, Hong Kong, p. 483 (2007)
21. Patwary, M.A., Palsetia, D., Agrawal, A., Liao, W.-K., Manne, F., Choudhary, A.: A new scalable parallel DBSCAN algorithm using the disjoint-set data structure. In: Proceedings of the International Conference on High Performance Computing, Networking, Storage and Analysis (SC'12), Salt Lake City, pp. 62:1–62:11. IEEE Computer Society, Los Alamitos (2012)
22. Roth, P.C.: ETRUSCA: event trace reduction using statistical data clustering analysis. Master's thesis, University of Iowa (1992)

23. Servat, H., et al.: Detailed performance analysis using coarse grain sampling. In: PROPER (2009)
24. Servat, H., et al.: Unveiling internal evolution of parallel application computation phases. In: ICPP, Taipei, pp. 155–164 (2011)
25. Shende, S.S., Malony, A.D.: The tau parallel performance system. Int. J. High Perform. Comput. Appl. **20**, 287–311 (2006)
26. Song, F., et al.: An algebra for cross-experiment performance analysis. In: ICPP, Montreal, pp. 63–72 (2004)
27. The CGPOP Miniapp website. http://www.cs.colostate.edu/hpc/cgpop
28. The libunwind project. http://www.nongnu.org/libunwind
29. The Weather Research & Forecasting model. http://www.wrf-model.org
30. Yilmaz, A., et al.: Object tracking: a survey. ACM Comput. Surv. **38**(4) (2006)

A Flexible Data Model to Support Multi-domain Performance Analysis

Martin Schulz, Abhinav Bhatele, David Böhme, Peer-Timo Bremer,
Todd Gamblin, Alfredo Gimenez, and Kate Isaacs

Abstract Performance data can be complex and potentially high dimensional. Further, it can be collected in multiple, independent domains. For example, one can measure code segments, hardware components, data structures, or an application's communication structure. Performance analysis and visualization tools require access to this data in an easy way and must be able to specify relationships and mappings between these domains in order to provide users with intuitive, actionable performance analysis results.

In this paper, we describe a data model that can represent such complex performance data, and we discuss how this model helps us to specify mappings between domains. We then apply this model to several use cases both for data acquisition and how it can be mapped into the model, and for performance analysis and how it can be used to gain insight into an application's performance.

1 Motivation

High Performance Computing (HPC) application developers are facing increasing complexity in supercomputer architectures as well as increasing complexity in the simulations that run on them. Modern HPC machines have deep memory hierarchies, complex network topologies, and accelerators such as GPUs and many-core chips. Applications use adaptively refined domain decompositions [1, 4, 9] and require complex coupling between disparate physics for scale-bridging. To understand these complexities and their interactions, developers must rely on tools for detailed performance information to optimize their codes. Moreover, measurements must be presented clearly and intuitively to enable actionable insights.

M. Schulz (✉) • A. Bhatele • D. Böhme • P.-T. Bremer • T. Gamblin
Lawrence Livermore National Laboratory, Livermore, CA, USA
e-mail: schulzm@llnl.gov

A. Gimenez • K. Isaacs
Lawrence Livermore National Laboratory, Livermore, CA, USA

University of California at Davis, Davis, CA, USA

© Springer International Publishing Switzerland 2015
C. Niethammer et al. (eds.), *Tools for High Performance Computing 2014*,
DOI 10.1007/978-3-319-16012-2_10

Performance tools have a bad track record in this respect. While they can collect a wide range of performance data and do so efficiently, the data reported by the tools is often very low-level and demands detailed system knowledge from the developer. Some tools, like Scalasca [23] or PerfExpert [3], try to close this gap by presenting higher level analysis results, but such tools are often limited by prior knowledge about potential bottlenecks that had to be coded into the tools; the detection of new bottlenecks or performance problems is often not possible, since it requires an in-depth understanding of code properties and performance data.

One fundamental problem with current tools is that data is displayed in a manner closely related to the way it was measured. While this is straightforward from a tool perspective, it often does not match the intuition of the developer and is hard to understand. For example, counts of cache misses or branch mispredictions are collected using hardware counters, then displayed on a per-core basis or, at best, mapped to source code. Their relationship to application physics, however, is left for users to infer.

We are addressing this issue with a data model that allows tool developers to (a) abstract measurements as values in independent data domains, (b) define mappings between domains to describe data transformations and (c) use data mapped into more intuitive domains for the visualization of performance information. We build on top of our previous work that proposed a simplified and limited three domain model [17] and we base our model on discussions at a recent Dagstuhl seminar, which covered the fundamentals of performance visualization [2]. This data model will enable developers to gain insights into the performance of complex systems using intuitive performance visualizations.

We are currently developing an architecture for multi-domain performance analysis, from data acquisition that is capable of gathering data from the entire software stack (including application level information), to data storage and queries, to novel visualization tools that utilize this information and are able to make use of the multi-domain nature of the data. In this paper we present three case studies: the use of hardware to simulation domain mappings, a tool to enable topological views of network data, and a tool to track data movements on NUMA systems. In all cases, the ability to map data from one domain into another for analysis and visualization is instrumental in extracting and understanding the insights necessary for performance optimization.

The remainder of this paper is organized as follows: Sect. 2 discusses how current tools collect and visualize data and how in some cases this does not match the intuition the user would like to see. We then briefly discuss our previous model [17] and its shortcomings. Based on these observations, we formulate the basics of a generalized model in Sect. 3 and then introduce an architecture to implement this model on large scale systems in Sect. 4. In Sect. 5 we present three case studies showing how the concept of cross-domain analysis can help in detecting performance problems, before we conclude in Sect. 6.

2 Tools and Their Data Domains

Performance analysis is a well established area in High Performance Computing (HPC) and many tools have been built and are in active use on HPC systems. Examples include Open|SpeedShop [16] and TAU [18], two general performance analysis frameworks; HPCToolkit [13], specializing in sampling based performance analysis; Vampir [15] and Jumpshot [24] for the analysis of communication traces; or mpiP [22] and ompP [7], two profilers for MPI and OpenMP communication respectively. These tools enable users to collect a wide variety of performance measurements based on timing information or using hardware performance counters exposing execution characteristics in the underlying system. In all cases, performance measurements are either tied to the physical hardware they were collected in or to execution objects, such as processes or threads, as defined by the programming model. We refer to this as the *measurement domain*.

While the information available is vast, it is also very low-level and requires substantial effort by a performance analyst to interpret the data and to convert the information into actionable insights. Because of this, many tools aim to automate performance analysis by detecting common bottlenecks, rather than by displaying raw performance information. The APART project [6] was among the first to formally define a catalogue of bottlenecks. Tools like Scalasca [23] or PerfExpert [3] use such information to search and identify predefined bottleneck patterns.

This approach, though, is naturally limited to existing, well-understood bottlenecks. While this is an important class of problems, new problems, including those caused by emerging architectures and applications, cannot be detected. Those will always require some manual analysis and will have to be done in close collaboration between a performance analyst and an application developer.

2.1 Collection vs. Analysis/Visualization Domain

For application developers, the information is not represented in an intuitive manner. One of the main reasons for this is that most tools deliver performance analysis results in the domains they were collected in. For example, hardware performance counters are shown per core ID or MPI message traffic is shown per MPI rank. Both are arbitrary spaces that have no intuitive meaning for the developer when taken alone. Developers are more familiar with different domains, like the application domain, e.g., the 2D or 3D domain of a physical phenomena that is being simulated, or the communication structure implemented on top of the MPI rank space.

While we cannot directly measure performance data in these intuitive domains, we can map data from a measurement domain into a (potentially) different visualization or analysis domain.[1] This decouples these two domains: data can be measured in arbitrary domains where it is available, while the user picks the (possibly different) domain in which she wants to see the data. This provides tools with a new flexibility that gives developers a novel way to interpret the data and thereby characterize the performance of their applications.

2.2 The HAC Model

The HAC (Hardware–Application–Communication) model, which we introduced in our previous work [17], was a first attempt to characterize and abstract this concept of decoupling the collection from the visualization domain. This model defined three domains, which we saw as the three most important domains, as well as mappings between them. This is illustrated in Fig. 1. The three domains were: H—the hardware domain, which can be used to represent the physical elements of a machine, like cores or network links, and which is used to measure hardware-related data like cache misses, floating point operations or number of packet sent over a link; A—the application domain in which the application's data resides, e.g., the simulated structure in a numerical simulation; and C—the communication domain,

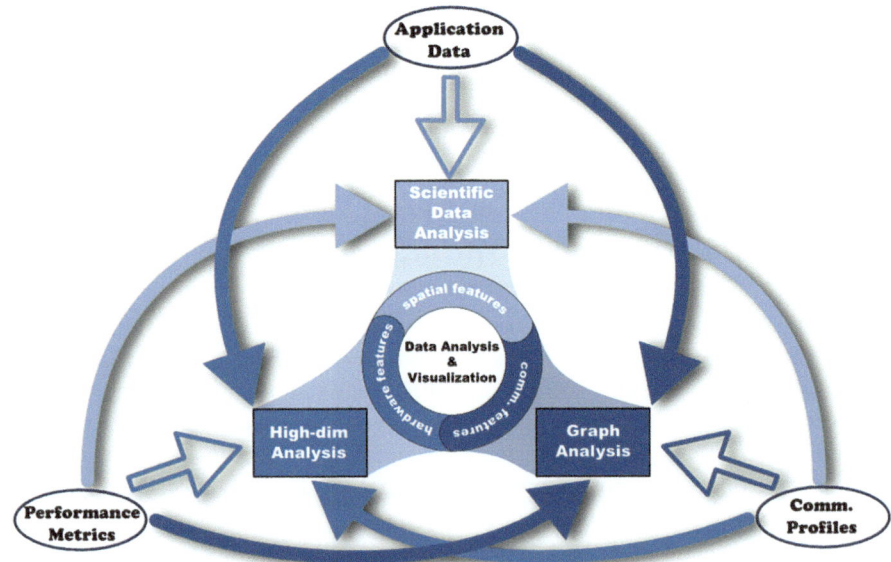

Fig. 1 Domains and mappings in the HAC model

---

[1] We will use these two terms interchangeably in the remainder of the paper.

which is used to abstract the communication between processes or threads within an application.

Each of these domains has its own properties and is associated with its own analysis and visualization techniques. Mappings between the domains can therefore help widen the number of analysis and visualization techniques on the data collected in any of the three domains and can make data collected in two different domains comparable. A straightforward example, which we will discuss in more detail in Sect. 5.1, is a mapping from the hardware to the application domain that can help attribute performance data collected on hardware resources to the sections of simulation data, which are computed by those resources.

2.3 Missing Elements in the HAC Model

While the HAC model provides the intended abstractions and enables us to provide users with the intended new insights, it only includes three very specific data domains. Information on data structures and memory distributions is not included, and the source code and time domains are handled separately and are not part of the model. Reasoning about them is a special case, limiting the scope of possible analyses.

Further, in HAC we treat mappings as static properties, which only holds for simple, non-adaptive cases. Codes with dynamic data and execution, such as Adaptive Mesh Refinement codes (AMR) [1, 4, 9], or dynamic environments, e.g., with thread migration among hardware threads or process migration for load balance optimization, however, need mappings that can be updated based on runtime events. This requires the online collection and integration of metadata.

3 A Generalized Data Model for Performance Tools

We present a generalization of the HAC model that overcomes the limitation described above. This model was the result of discussions at a Dagstuhl seminar on Performance Visualization [2] and has input from both the performance analysis and visualization communities. Figure 2 provides a high-level sketch of the concepts explained below.

3.1 Spaces and Domains

At the core of the abstraction is a set of spaces. Each space is represented by a finite set of tuples and has a cross product of types associated with it, such that each type describes one element of the tuple. The number of spaces is not limited. Time and

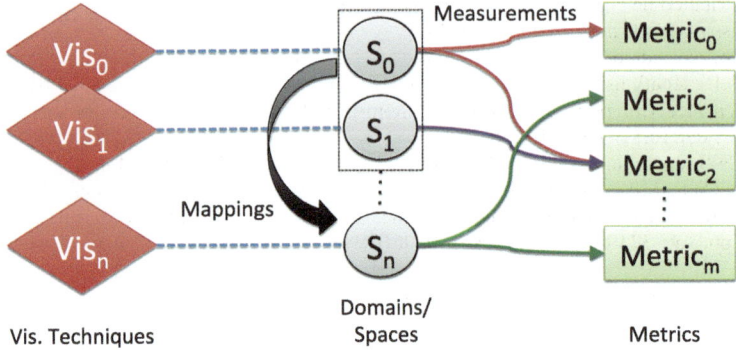

Fig. 2 A generic data model to capture relationships between domains

code (represented by calling context trees [20]) are treated the same as any other space. A domain is represented by a cross product of spaces.

3.2 Metrics

Metrics are units for individual data points. Examples are floating point operations and MPI message counts. Metrics are typically represented by infinite sets, as not to restrict what can be measured, but may in individual cases be a finite set of possible outcomes.

3.3 Measurements

Measurements capture the data acquisition in performance tools. They are represented as mappings of a cross product of spaces, the domain the performance data is collected in, to a metric, the set of possible values for this measurement. To make reasoning about measurements easier, we define a measurement as a unique mapping or function, i.e., for each element of the measurement domain the measurement only maps to at most one element in the metric set. If this is not the case for an experiment, e.g., in profiling tools that provide multiple data points for each element of a space over time, the domain needs to be modified to allow for this uniqueness, in the example by adding a space representing real or virtual time to the cross product that forms the domain.

3.4 Mappings

A mapping, in the sense of the data model, maps one or more spaces (the origin domain) to one or more spaces (the target domain). This allows measurements represented in the target domain to be used in analysis operations on the origin domain. In general, we can distinguish three types of mappings, which are also illustrated in Fig. 3:

- **1:1 Mappings:** each element of the origin domain is mapped to exactly one element of the target domain. An example of such a 1:1 mapping is the mapping between node coordinates in a network to node IDs, since both domains describe the same physical entity, but using different names or numbering schemes. 1:1 mappings allow a direct translation of measurements in one domain to another.
- **1:N Mappings:** each element of the origin domain is mapped to one or more elements of the target domain. An example for a such a 1:N mapping is the mapping from node IDs in a system to process or MPI rank, since multiple ranks can be on each node. When mapping measurements using a 1:N mapping, measurements from all elements in target domain that map to a single element in the origin domain have to be combined using an aggregation operation. This can be as simple as a sum or average, but can also be a more complex operation such as clustering or statistical outlier detection.
- **N:1 Mappings:** each element of the origin domain is mapped to at most one, not necessarily unique, element of the target domain. An example for a such a N:1 mapping is the mapping of MPI ranks to nodes in a system, since multiple ranks can be on each node. When mapping measurements using a N:1 mapping, a measurement from an element in the target domain must be distributed or spread over all elements in the origin domain that map to it. The semantics of this operation depends on the semantics of the domains. For example, the same measured value could be attributed to each element in the origin domain in full, or the value could be split up based on a distribution function.

Mappings can further be combined into new mappings, allowing a translation over multiple domains from an origin to a target domain. This could also lead to situations in which multiple translations between two domains using different compositions, i.e., a different route through the set of available domains, are

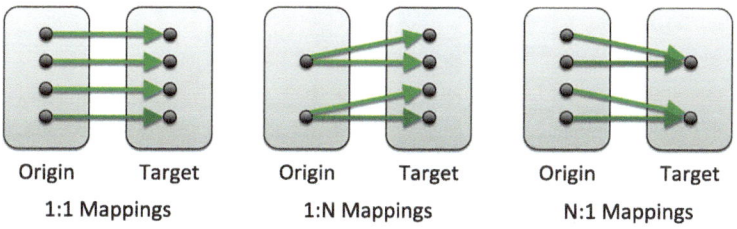

Fig. 3 Types of mappings

possible. Note, though, that not all combined mappings between the same domains carry the same semantics. For example, a 1:1 mapping between two domains may also be representable by a combination of a N:1 and a 1:N mapping, but the latter would include a loss of information by first aggregating measured values before spreading them out again. Choosing the right combination of mappings based on the intended analysis is therefore crucial.

4   An Architecture to Enable Cross-Domain Analysis

To implement this model, we require a performance analysis pipeline that allows us to not only collect performance data, but also collect the necessary context to establish the mappings between spaces. Both should then be stored in a scalable data store that offers a flexible query API to tools so they can extract the data based on the intended visualization domain. We are currently developing such an infrastructure and Fig. 4 shows its high-level architecture.

4.1   Software Stack Instrumentation

To extract the necessary context that allows us to establish mappings between spaces, we need access to information from the entire software stack, including from the OS and adaptation decisions it makes, and from runtime systems and their abstractions established for application programmers. For this purpose we need access to the necessary interfaces, such as the /proc/ interface in Linux,

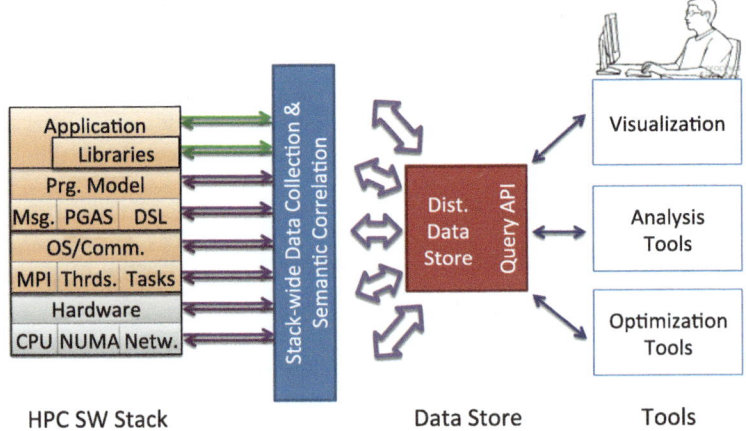

Fig. 4 An architecture for multi domain performance analysis

or access to machine specific instructions and APIs providing us with core and node information. On the programming model side we need standardized mechanism to bridge the abstraction provided by the model. For MPI, the standardized interfaces PMPI (the MPI Profiling Interface) and MPI_T, the newly defined MPI tool information interface [14], can provide the necessary information, but other programming models and runtime systems do not provide the necessary information through a standardized interface.

Over the last year, though, the tools working group in the OpenMP language committee has been developing a new interface that will allow tools to interface with any OpenMP runtime and export the information necessary. This interface, OMPT [5], provides a series of routines to extract runtime information, e.g., to cleanly assemble callstacks, and to insert hooks for events of interest, such as the start of parallel regions or tasks. Initial prototype implementations of OMPT are available on BlueGene/Q machines and Intel platforms. We are currently integrating OMPT into our tools.

4.2 Creating Context

In many cases, though, information provided transparently by runtimes is not sufficient to provide all metadata needed to establish mappings. Many kinds of information are application specific and we need application semantics to properly represent, capture, and store them. Examples are program phases, associations of tasks and data structures, or application specific properties of an input deck. We therefore need an API that enables developers to expose this information in an unobtrusive and tool-independent way.

For this purpose, we have developed a context recording library that gives developers simple commands to annotate application source code and provide context information through key/value pairs. The library then records this context information and makes it available to performance data collection tools. The latter is done by providing a reference to a context information structure, which allows the context recording library to maintain a highly efficient internal representation of the data. At program termination, the library then writes this context information to disk, which later can be extracted during the analysis phase.

Figure 5 shows a few code snippets illustrating how context annotation calls are added to the application source code. In this example, one context key/value pair consists of key `iteration`, which expresses the current iteration that the execution of the target code is in. Developers can also define hierarchical annotations, as shown with the `phase` annotation in the example.

```
1   Annotation("phase").begin("main");  // phase="main"
2
3   Annotation("phase").begin("loop");  // phase="main/loop"
4   while ( i < count ) {
5     Annotation("iteration").set(i);
6     do_something( i );
7   }
8   Annotation("phase").end();  // ends "main/loop"
9   Write_results();
10  Annotation("phase").end();  // ends "main"
```

Fig. 5 Example of source code context annotations

4.3 A Distributed Data Store with Query Access

Once collected, data and metadata will be stored in a large and distributed data store
to make it available to tools, enabling queries across different data sources, runs,
systems and applications. Our initial implementation used for Boxfish [12] relies on
simple, self-describing text files, which capture information from multiple domains
with a column per measured metric. This data format has generally been very
effective in order to achieve the necessary cross-space connections, but naturally
has its limits in terms of scalability and performance. We are therefore currently
analyzing both SQL and non-SQL based stores as possible alternatives.

One key aspect of the data store will be that it can execute complex queries, which
requires it to track existing mappings and apply them in a suitable order to achieve
the necessary transformations. The latter includes finding the right combination
of mappings as well as the scheduling of necessary aggregation and spreading
functions. If multiple chains of mapping between targets and sources are available
and are feasible, the query API should first select a set of likely paths, e.g., ones
without huge amounts of data loss due to combining N:1 and 1:N mappings, and
then, if ambiguity persists, should forward the decision to the user.

4.4 Analysis and Visualization Tools

From this data store, tools can query performance data based on the visualization
domain they prefer or need for their analysis. Generally speaking, we envision three
types of tools: (a) performance visualization tools, such as Ravel [11], Boxfish [12]
and communication visualizations as the one by Sigovan et al. [19], (b) analysis
tools that preprocess the data and only deliver reduced performance analysis results,
or (c) tools that can autonomously use the information for tuning, such as Active
Harmony [21].

5 Case Studies

In the following we show the advantages and capabilities of our multi-domain approach using three case studies.

5.1 Hardware to Simulation Domain Mappings

To demonstrate how our new abstract data model described above can be instantiated, we first show a straightforward example from a CFD simulation at LLNL, which we previously analyzed using the HAC model [17], and show how it can be represented in the new, more general model. In this experiment, we simulate an ablator driving a shock into an aluminum section containing a void, producing a jet of aluminum. The simulation results are from a 12 h run using a 2D grid running on 256 (8 × 32) cores of a Linux based cluster with nodes consisting of four Quad-Core AMD processors each. The interconnecting network is Infiniband and we use the MVAPICH MPI implementation. Figure 6b, c show the results of the simulation using nine selected timesteps in regular intervals sorted from left to right. The figures clearly show the aluminum jet on the left side as well as the created shockwave traveling from top to bottom.

Figure 6a shows the instantiation of the data model for this experiment: in total we have three domains: the 2D application domain (defined by two spaces for X and Y coordinates) representing the physical structure being simulated, in this case the aluminum shock wave; the MPI rank domain (defined by a single space holding the rank ID), in which most tools operated; and the hardware domain (defined by three spaces, one for each dimension of the machine structure), describing the structure of the machine and its node architecture. During the experiment we collect metrics both in MPI rank space (number of floating point operations and number of L1 cache misses) and two physical properties of the shockwave (the material density itself and the material velocity).

Each of the three domains has its own visualization technique associated with it: the MPI rank space is represented by a simple bar graph (Fig. 6f shows the number of floating point operations natively collected in this domain), the application space is shown using a 2D scientific visualization (Fig. 6b, c show the two metrics natively collected in this domain); and a representation of the hardware domain with nodes, sockets, and cores is shown in Fig. 6g. Just looking at these native representations, the performance data collected in MPI rank space cannot easily be correlated with any other measurement, which makes it hard to interpret the variations in floating point operations that we observe.

Using mappings between the domains, we can establish such a correlation: based on our machine and MPI setup, we can assume a fixed mapping of MPI ranks to cores and therefore we can map the MPI rank domain into the hardware domain (and vice versa) using a static 1:1 mapping. We can further map the rank domain

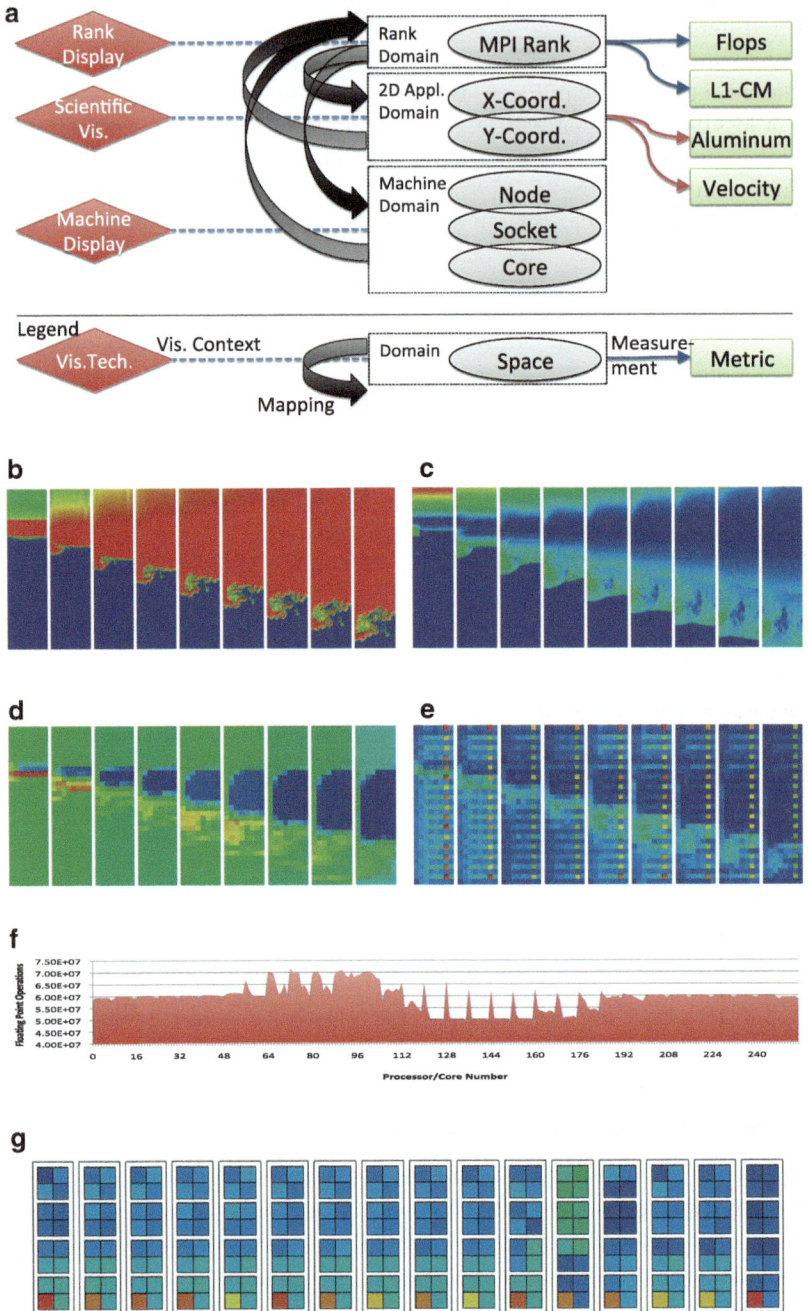

Fig. 6 Example instantiation of the data model using a CFD code. (a) Abstract representation using the generalized data model. (b) Aluminum in 2D scientific vis. (c) Velocity in 2D scientific vis. (d) Flop/s in 2D scientific vis. (e) Cache Misses in 2D scientific vis. (f) Flop/s, measured per rank, displayed per rank. (g) Cache Misses display in machine topology

into the application domain by following the domain decomposition: an MPI rank is mapped to all elements in the application domain it is responsible for computing. In this case, the domain decomposition splits up a dense matrix in a 8x32 grid, making each MPI rank responsible for one of 256 sub-squares of the full problem. Since there are more data points in the application domain, the mapping from MPI rank domain to the application is a 1:N mapping, while the reverse direction is a N:1 mapping.

Figure 6d, e show the two metrics measured in the MPI rank domain mapped into the 2D application domain and shown using the matching visualization technique. Since multiple elements in the application domain map to a single MPI rank, the associated N:1 mapping from the application domain to the rank domain spreads the measurements of a single rank to all elements in application domain, resulting in a more coarse-grained representation. Nevertheless, this displays shows a clear correlation with the simulated physics in Fig. 6b, c for both metrics, which wasn't obvious before in the MPI rank display.

The visualization of the L1 cache misses in Fig. 6e shows a second phenomenon: a series of ranks with high numbers of cache misses represented by dots on the right side of the figure for each timestep, which overlays the physics, but does not seem to be correlated. When we map the same data into the hardware domain (Fig. 6g) we see that each of these cores maps to the same core ID on each node, and we tracked this phenomena to the MPI library, which uses this core for aggregating node local information for collective operations.

Using the two different mappings into two domains enabled us to distinguish these two separate phenomena and what they correlate to. It also showed that they need to be analyzed separately since they have two different sources: one related to the physics being simulated, i.e., input dependent, and one related to the base architecture, i.e., input independent.

5.2   Domain Mappings in Boxfish

In order to automate the analysis process described above, we have implemented a prototype of the query and mapping functionality in the Boxfish toolset [12]. An annotated screenshot of our tools with example mappings is shown in Fig. 7.

The tool offers a range of visualization techniques, shown in the bottom left of the figure. Our current implementation focuses mainly on visualization of communication data, but it can be extended with plugins to other domains. The input data and its metrics are shown in the top left window in the figure. Each *table* represents one measurement domain and the elements below it show the various metrics available in that domain.

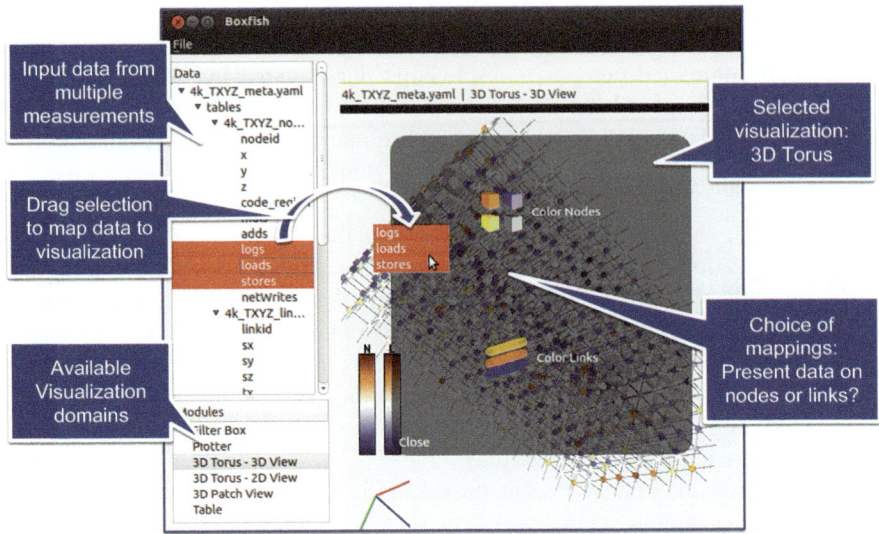

Fig. 7 Multidomain visualization in Boxfish

A user can select any visualization domain by dragging it into the main window. The example shows a 3D visualization of a torus network, as used in Blue Gene/L and P as well as Cray HPC systems. Once the display is activated, the user can select any number of metrics available as input data and attempt a mapping of this data onto the visualization domain, again by simply dragging the metric onto the display. At that time, Boxfish searches all available mappings, provided through plugins into the tool, to see if a mapping from the selected visualization to the selected input metrics and domain can be established. If the mapping is unique, the data is simply displayed accordingly. If multiple mappings are available, though, the tool presents this choice to the user. In the example in the figure, the user drags a node metric onto the torus display. However, since the tool also knows how to map node IDs to link IDs (since this is how link data is measured on Blue Gene systems), Boxfish offers the user the choice of visualizing the data on nodes or links. The user picks the intended one by dragging the data onto that choice.

While the choice is clear to a human user in this case, the system cannot distinguish between these two choices without further semantic information. This shows the generality of the approach, but also a shortcoming. We will investigate how to add semantic information to reduce choices offered by Boxfish to reasonable mappings in future work. In other examples, though, several choices do make sense (e.g., attributing packets sent over a link can be to a link to show network traffic vs. to a node to show network load on a node). In these cases the system provides users with the necessary flexibility to choose the option suitable for the intended analysis.

5.3   NUMA Optimizations with MemAxes

In addition to Boxfish, we started using the concept of domain mappings in other tools. One example is MemAxes [8], a tool to visualize memory traffic in NUMA nodes. It uses Intel's Precise Event Based Sampling (PEBS) to sample all memory accesses and their attributes, including latency, target address and memory hierarchy they hit. This information is further enhanced with context information containing both source code meta data (e.g., which element in which structure an address maps to) and machine information (e.g., core ID to NUMA domain mappings). Figure 8 shows an overview of the graphical user interface. The elements labeled with A show the application view of the collected data, mapping it to data structures and source code. The view labeled B shows a mapping of the collected data to the hardware architecture of the underlying system: HW threads are aligned on the outer most ring, following by the—partially shared—caches all the way to the two NUMA domains in the innermost circle. The amount of communication between the layers is shown through the thickness of the lines between the layers. Users can select only parts of the data being shown, which updates the various views in the GUI. The graph in window labeled with C shows the percentage of the data that is currently shown.

The selection can be done using a parallel coordinates view in the bottom most window D. Each vertical bar (axis) represents one measurement on one metric (across all spaces in which data is collected) and a single sample is drawn as a line between all axes showing the value in that particular measurement for this sample. Users can select arbitrary ranges on any of the shown measurements by marking a region of the axis, which will then color all lines crossing the axis at that

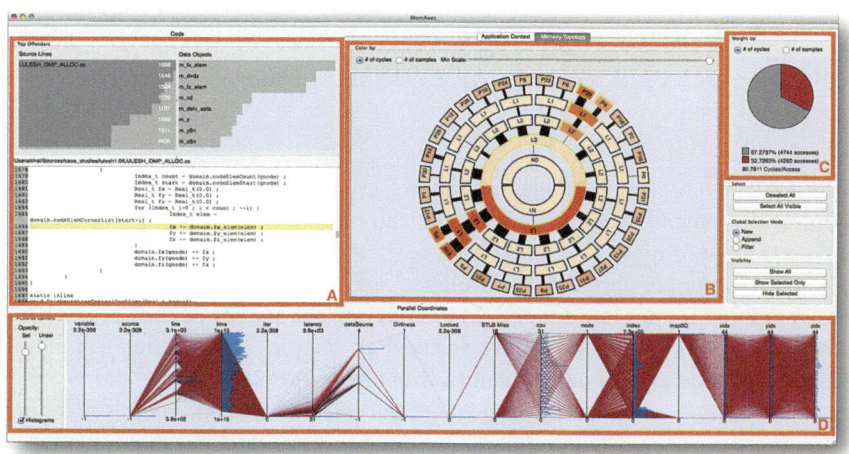

Fig. 8 The Memaxis user interface

a

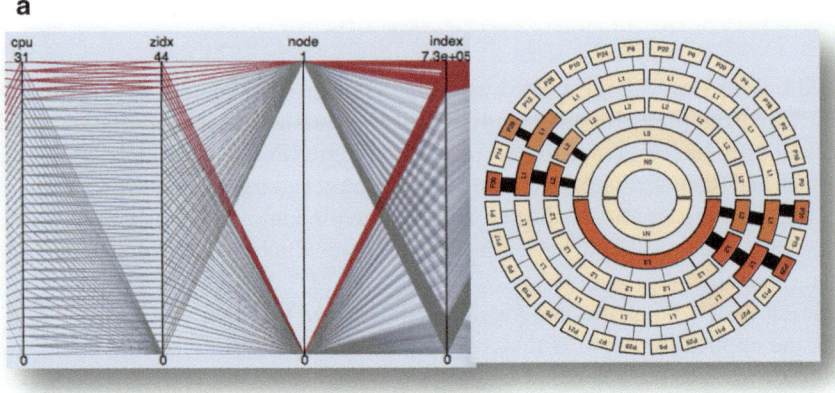

b

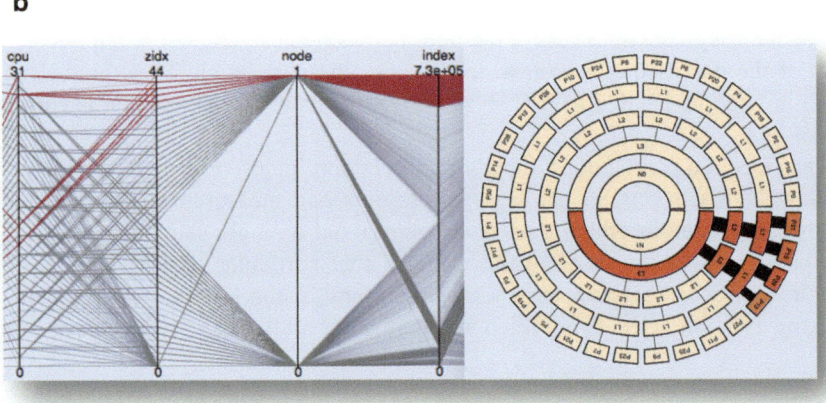

Fig. 9 Using MemAxes to expose memory traffic in NUMA systems. (**a**) Initial version of LULESH: memory accesses across NUMA domains. (**b**) LULESH after memory access optimizations

measurement red. This visualization shows how other measurements and spaces correlate to elements being selected in the chosen space.

An example of how this can be used is shown in Fig. 9: We ran LULESH [10], a shock-hyrdo proxy application developed at LLNL, on a NUMA system with 32 hardware threads split into two NUMA domains. The results of the default run are shown in Fig. 9a. Using the tool, we selected all samples representing a continuous range in the code's main data structure (right bar) and observed the corresponding memory traffic. As the architectural view clearly shows, the contiguous mapping of this data structure matches the contiguous set of core IDs, but those are mapped in alternating order to both NUMA nodes (second bar from the right). As a consequence, the system experiences significant traffic between NUMA regions.

To optimize this behavior we rearrange the cores by alternating core IDs, shown in Fig. 9b. As a result, consecutive ranges in the main data structure are now mapped to core IDs in the same way they are in the NUMA domain. This lead to a performance improvement of over 10 %.

6   Conclusions

Performance analysis plays a critical role in optimizing codes on current and future platforms. The ever increasing complexity of both system architectures and applications makes this a difficult task and application developers rely on tools to provide them with sufficient insights into the performance behavior of their applications. While many tools exist and can provide a vast amount of data, the resulting information is often low-level and hard to interpret. One reason for this is that information is displayed in domains in which data was collected, but those domains are not necessarily the intuitive ones that help application developers extract actionable insights.

In this work we introduced a generic data model that helps us describe independent data domains for both the collection of measurements and their visualization/analysis. By establishing mappings between domains we can then translate data collected in one domain into another and use this to show performance data in other domains, to make data comparable across domains, and to provide intuitive visualizations in domains that application developers are familiar with. This concept has already been helpful in many cases and we showed three of them in this paper: mapping of performance data to the application domain of a CfD simulation helped us correlate performance data with the underlying physics being simulated and to distinguish those effects from a machine related phenomena; the Boxfish tool enabled us to understand network performance data by mapping it to the underlying physical network architecture, in our case a 3D torus; and correlating memory access information from the machine architecture and the application data structures allowed us to identify and correct excessive NUMA accesses for a shock-hydro code.

In summary, the new data model and its ability to map performance data across domains enables us to create a new generation of tools that provide more insights into an application's performance and provide this information in an intuitive way that enables optimizations. It is also flexible enough to carry forward to next generation systems and applications, including new programming models and abstractions, by integrating more diverse domains and measurement techniques. Equally important, it provides a way to formalize performance data to enable a closer interaction between the performance analysis community on one side and the data visualization and analysis communities on the other, which will allow for an easier transfer of tools and techniques.

Acknowledgements This work was performed under the auspices of the U.S. Department of Energy by Lawrence Livermore National Laboratory under Contract DE-AC52-07NA27344 (LLNL-CONF-664263) and supported by Office of Science, Office of Advanced Scientific Computing Research as well as the Advanced Simulation and Computing (ASC) program.

References

1. Bell, J., Almgren, A., Beckner, V., Day, M., Lijewski, M., Nonaka, A., Zhang, W.: BoxLib user's guide. https://ccse.lbl.gov/BoxLib/BoxLibUsersGuide.pdf (2013)
2. Bremer, P.-T., Mohr, B., Pascucci, V., Schulz, M.: Connecting performance analysis and visualization to advance extreme scale computing. http://www.dagstuhl.de/de/programm/kalender/semhp/?semnr=14022 (2014)
3. Burtscher, M., Kim, B.-D., Diamond, J., McCalpin, J., Koesterke, L., Browne, J.: Perfexpert: an easy-to-use performance diagnosis tool for HPC applications. In: Proceedings of the 2010 ACM/IEEE International Conference for High Performance Computing, Networking, Storage and Analysis (SC'10), Washington, DC, pp. 1–11. IEEE Computer Society (2010)
4. Colella, P., Graves, D.T., Keen, N.D., Ligocki, T.J., Martin, D.F., McCorquodale, P.W., Modiano, D., Schwartz, P.O., Sternberg, T.D., Van Straalen, B.: Chombo software package for AMR applications design document. Technical report, Applied Numerical Algorithms Group, Computational Research Division, Lawrence Berkeley National Laboratory, Berkeley, 15 Apr 2009
5. Eichenberger, A.E., Mellor-Crummey, J., Schulz, M., Wong, M., Copty, N., Dietrich, R., Liu, X., Loh, E., Lorenz, D.: OMPT: an OpenMP tools application programming interface for performance analysis. In: Rendell, A.P., Chapman, B.M., Mller, M.S. (eds.) OpenMP in the Era of Low Power Devices and Accelerators. Number 8122 in Lecture Notes in Computer Science, pp. 171–185. Springer, Berlin/Heidelberg (2013)
6. Fahringer, T., Gerndt, M., Riley, G., Traff, J.L.: The APART Specification Language (illustrated with MPI). Technical report FZJ-ZAM-IB-2001-08, Forschungszentrum Jülich (2001)
7. Fürlinger, K., Gerndt, M.: ompP: A profiling tool for OpenMP. In: Proceedings of the First International Workshop on OpenMP (IWOMP'05), Eugene, 1–4 June 2005
8. Giménez, A., Gamblin, T., Rountree, B., Bhatele, A., Jusufi, I., Bremer, P.-T., Hamann, B.: Dissecting on-node memory access performance: a semantic approach. In: Proceedings of the ACM/IEEE International Conference for High Performance Computing, Networking, Storage and Analysis (SC'14), New Orleans, LA. IEEE Computer Society (2014). LLNL-CONF-658626
9. Hornung, R.D., Kohn, S.R.: Managing application complexity in the SAMRAI object-oriented framework. Concurr. Comput. Pract. Exp. **14**(5), 347–368 (2002)
10. Hydrodynamics challenge problem, lawrence livermore national laboratory. Technical report LLNL-TR-490254
11. Isaacs, K., Bremer, P.-T., Jusufi, I., Gamblin, T., Bhatele, A., Schulz, M., Hamann, B.: Combing the communication hairball: visualizing large-scale parallel execution traces using logical time. In: Proceedings of IEEE Symposium on Information Visualization (InfoVis'14), Paris, France (2014)
12. Landge, A.G., Levine, J.A., Isaacs, K.E., Bhatele, A., Gamblin, T., Schulz, M., Langer, S.H., Bremer, P.-T., Pascucci, V.: Visualizing network traffic to understand the performance of massively parallel simulations. In: Proceedings of IEEE Symposium on Information Visualization (InfoVis'12), Seattle, 14–19 Oct 2012. LLNL-CONF-543359
13. Mellor-Crummey, J., Fowler, R., Marin, G.: HPCView: a tool for top-down analysis of node performance. J. Supercomput. **23**, 81–101 (2002)
14. MPI Standard 3.0. http://www.mpi-forum.org/docs/docs.html

15. Nagel, W.E., Arnold, A., Weber, M., Hoppe, H.C., Solchenbach, K.: VAMPIR: visualization and analysis of MPI resources. Supercomputer **12**(1), 69–80 (1996)
16. Schulz, M., Galarowicz, J., Maghrak, D., Hachfeld, W., Montoya, D., Cranford, S.: Open|speedshop: an open source infrastructure for parallel performance analysis. Sci. Program. **16**(2–3), 105–121 (2008)
17. Schulz, M., Levine, J.A., Bremer, P.-T., Gamblin, T., Pascucci, V.: Interpreting performance data across intuitive domains. In: 2011 International Conference on Parallel Processing (ICPP), Taipei City, pp. 206–215 (2011)
18. Shende, S., Malony, A.D.: The tau parallel performance system. Int. J. High Perform. Comput. Appl., ACTS Collection Special Issue **20**(2), 287–311 (2005)
19. Sigovan, C., Muelder, C.W., Ma, K.-L.: Visualizing large-scale parallel communication traces using a particle animation technique. Comput. Graph. Forum **32**(3), 141–150 (2013)
20. Tallent, N.R., Adhianto, L., Mellor-Crummey, J.M.: Scalable identification of load imbalance in parallel executions using call path profiles. In: Proceedings of IEEE/ACM Supercomputing'10, New Orleans (2010)
21. Tapus, C., Chung, I.-H., Hollingsworth, J.K.: Active harmony: towards automated performance tuning. In: Proceedings of the Conference on High Performance Networking and Computing, Phoenix, pp. 1–11 (2003)
22. Vetter, J., Chambreau, C.: mpiP: Lightweight, Scalable MPI Profiling. http://mpip.sourceforge.net, March 13, 2014
23. Wolf, F., Wylie, B., Abraham, E., Becker, D., Frings, W., Fuerlinger, K., Geimer, M., Hermanns, M.-A., Mohr, B., Moore, S., Szebenyi, Z.: Usage of the SCALASCA toolset for scalable performance analysis of large-scale parallel applications. In: Proceedings of the 2nd HLRS Parallel Tools Workshop, Stuttgart (2008)
24. Zaki, O., Lusk, E., Gropp, W., Swider, D.: Toward scalable performance visualization with jumpshot. Int. J. High Perform. Comput. Appl. **13**(3), 277–288 (1999)